2D NMR

Density Matrix
and
Product Operator Treatment

Gheorghe D. Mateescu
and
Adrian Valeriu

Department of Chemistry
Case Western Reserve University
Cleveland, Ohio

A SOLOMON PRESS BOOK

P T R Prentice Hall, Englewood Cliffs, New Jersey 07632

Library of Congress Cataloging-in-Publication Data

Mateescu, Gh. D.
2D NMR : density matrix and product operator treatment / G.D.
Mateescu and A. Valeriu.
p. c.
"A Solomon Press book."
Includes bibliographical references and index.
ISBN 0-13-013368-X
1. Nuclear magnetic resonance--Technique.. 2. Density matrices.
3. Product formulas (Operator theory) I. Valeriu, A. .
II. Title.
QC762.M267 1993
538'.362--dc20 92-32291
CIP

Editorial/production supervision and interior design: The Solomon Press
PTR Production Liason: Mary Rottino
Acquisitions editor: Betty Sun
Cover design: The Solomon Press
Manufacturing buyer: Margaret Rizzi

The publisher offers discounts on this book when
ordered in bulk quantities. For more information,contact:

Corporate Sales Department
PTR Prentice Hall
113 Sylvan Avenue
Englewood Cliffs, New Jersey, 07632

Phone: 201-592-2863
Fax: 201-592-2249

A Solomon Press Book

Printed in the United States of America
10 9 8 7 6 5 4 3 2 1

ISBN 0-13-013368-X

Prentice-Hall International (UK) Limited, *London*
Prentice-Hall of Australia Pty. Limited, *Sydney*
Prentice-Hall Canada, Inc., *Toronto*
Prentice-Hall Hispanoamericana, S.A., *Mexico*
Prentice-Hall of India Private Limited, *New Delhi*
Prentice-Hall of Japan, Inc., *Tokyo*
Simon & Schuster Asia Pte. Ltd., *Singapore*
Editoria Prentice-Hall do Brasil, Ltda., *Rio de Janeiro*

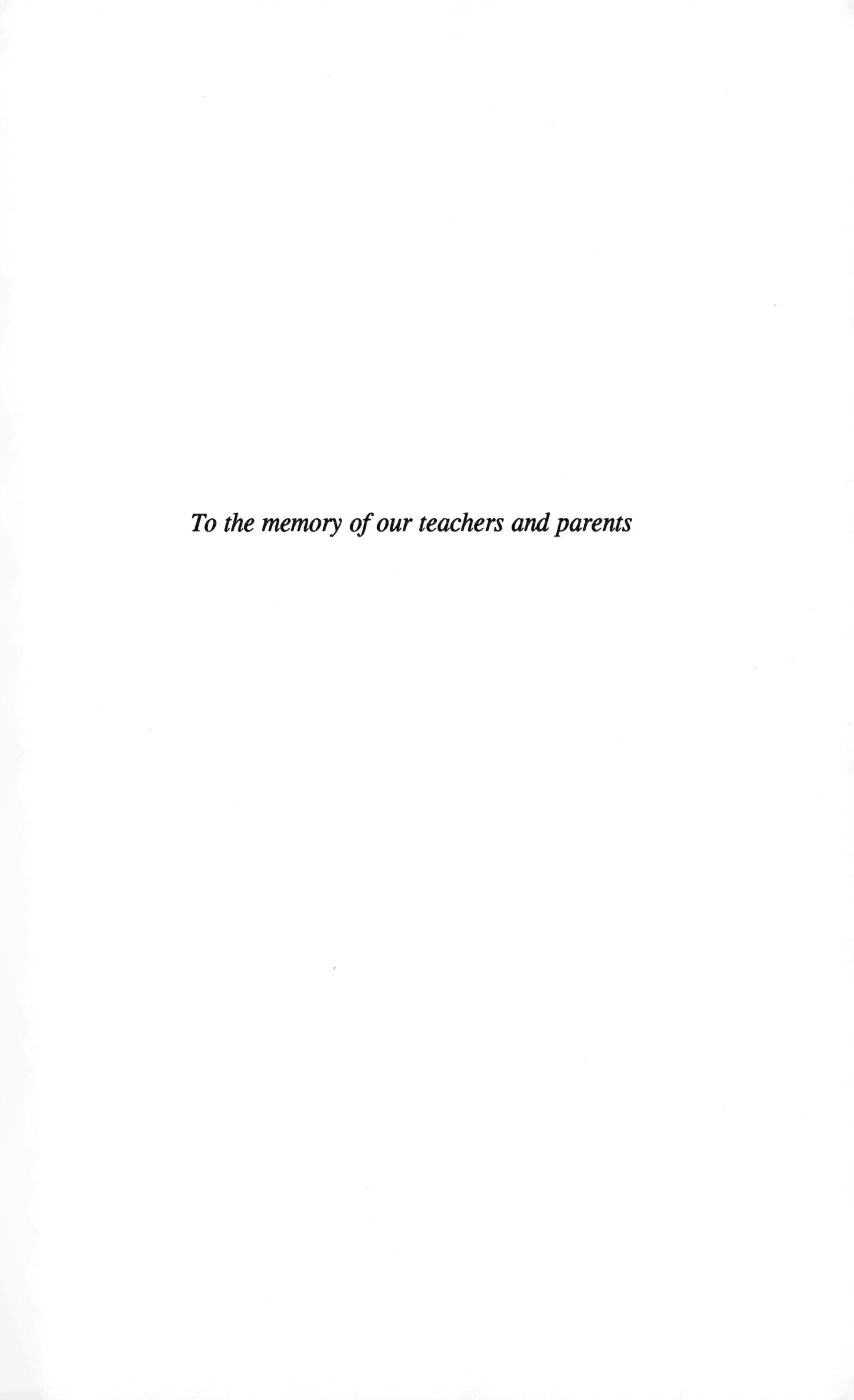

To the memory of our teachers and parents

Contents

Appendices

Foreword

There is hardly any doubt that NMR is nowadays the most powerful and perhaps also the most beautiful analytical technique. Its versatility and its range of applications is truly unlimited.

NMR is unique in the sense that it derives its power largely from a quantum mechanical understanding of its foundations. Those who master a mathematical description of NMR experiments have an enormous advantage over users who blindly follow the rules of the instruction manuals.

The most elegant, most simple, and also most intuitive description developed so far is the product operator formalism that is the central subject treated in this book. This formalism is like a magic key that provides access to the enormous arsenal of NMR techniques available today and permits the user to select and properly apply the most suited tools as well as to develop himself novel, perhaps even more useful, techniques.

I am convinced that this pedagogically very well done book by Gheorghe D. Mateescu and Adrian Valeriu will serve its purpose exceptionally well in the hands of numerous novices that intend to enter this fascinating field of science.

Richard R. Ernst

Asilomar, April 1992

Preface

Until recently, the teaching and understanding of modern (pulse) Nuclear Magnetic Resonance has made successful use of vector descriptions, including handwaving, since the pulse sequences were relatively simple. The advent of two-dimensional NMR made it practically impossible to explain the intricate effects of combined pulses and evolutions exclusively on the basis of vector representation. It thus became necessary to use an appropriate tool, the density matrix (DM) formalism. The DM treatment is generally found in specialized books which emphasize its quantum mechanical foundation. The quantum mechanical approach, however, constitutes a significant barrier for a growing number of students and scientists in the fields of chemistry, biology, medicine and materials research who want to gain a better understanding of 2D NMR.

This book constitutes a guide for the use of density matrix calculations in the description of multipulse NMR experiments. In keeping with its didactic nature, the text follows a step-by-step procedure which contains more detail than usual. This will give readers with modest mathematical background the possibility to work out or to create sequences of various degrees of complexity. Our treatment begins with an intuitive representation of the density matrix and continues with matrix calculations without trying to explain the quantum mechanical origin of pulse effects (*rotations*) and *evolution* of the matrix elements. The quantum mechanical approach is deferred to Appendix B. Those who do not want to take anything for granted may actually begin with Appendix B (it is assumed, of course, that the reader is familiar with the principles and experimental aspects of Fourier transform NMR).

The first part of the book contains a detailed DM description of the popular two-dimensional sequence, 2DHETCOR (2D heteronuclear correlation). It starts with the characterization of the system of nuclei at equilibrium in a magnetic field and concludes with the calculated signal which results from application of pulses and evolutions. This section is written in such a way as to be accessible to students with only an undergraduate mathematical background (there is even a Math Reminder in Appendix A). In order to ensure the continous flow of the minimal information needed to understand the sequence without too many sidetracks a number of detailed calculations of secondary importance are given in Appendix I.

Once familiar with 2DHETCOR, the student is led, step-by-step, through the calculations of a double-quantum coherence sequence and those of the widely used COSY (correlation spectroscopy). Throughout this book we did not use the t_1 and t_2 notations for the two time variables in a 2D experiment.in order to avoid confusion with the relaxation times T_1 and T_2. Also, there is still no consensus as to what notation should be given to the detection period (some call it t_1, some t_2).

The second part is entirely dedicated to the product operator (PO) formalism. The student will appreciate the significant economy of time provided by this elegant condensation of the density matrix procedure. He or she will be able to handle in reasonable time and space systems of more than two nuclei which would require much more elaborate calculations *via* the unabridged DM treatment.

Appendix B offers an accessible quantum mechanical presentation of the density matrix. Appendix C contains a selection of angular momenta and rotation operators written in matrix form, while Appendix D summarizes the properties of product operators. Appendices E through M are for students interested in a demonstration of the relations and procedures used in the text.

Throughout the book, relaxation processes have been neglected; this does not affect the essential features of the calculated 2D spectra and contributes to the clarity of the presentation.

The teaching method presented in this book has been successfully used in an Instrumental Analytical Chemistry graduate course for the past few years at Case Western Reserve and in several short courses. Being essentially a self-sufficient teaching tool (lecture notes), this book does not contain literature references. Numerous citations can be found in the books indicated in the Suggested Readings section. One of us being a passionate skier, we may say our *class* is for *beginners,* Farrar and Harriman's, for *intermediates,* and Ernst-Bodenhausen-Wokaun's, for *advanced.* In fact, our work is a synergic complement to Martin and Zektzer's *Two-Dimensional NMR Methods for Establishing Molecular Connectivity: A Chemist's Guide to Experiment Selection, Performance, and Interpretation.*

Acknowledgments

We thank Guy Pouzard and Larry Werbelow of the Université de Provence (Marseille) for stimulating discussions which inspired our endeavor. The first presentations of our way of teaching the density matrix formalism were at two NATO Advanced Study Institutes organized in 1983 by Camille Sandorfy and Theophile Theophanides in Italy and by Leonidas Petrakis and Jacques Fraissard in Greece. We thank for permission to use material from those papers, published by D. Reidel Publishing Company (listed in the Suggested Reading Section). We also thank Frank Anet of UCLA and Gary Martin of the Burroughs Welcome Co. for their comments and encouragement. Our publishing team has done an excellent job and we wish to thank Sidney Solomon and Raymond Solomon of the Solomon Press, Betty Sun of Prentice Hall, and the wonderful copyeditor Mary Russell. Last, but not least, we acknowledge the loving care and understanding of our spouses, Claudia and Anca.

1

The Density Matrix Formalism

1. INTRODUCTION

Only the simplest NMR pulse sequences can be properly described and understood with the help of the vector representation (or handwaving) alone. All two-dimensional experiments require the density matrix formalism. Even some one-dimensional NMR sequences (see Part II.12) defy the vector treatment because this approach cannot account for the polarization transfer. The goal of Part I is to show how the density matrix can be used to understand a specific NMR pulse sequence. A "math reminder" is given in Appendix A for those who may need it. After becoming familiar with the use of the density matrix as a tool, the reader may find enough motivation to go to Appendix B which deals with the quantum-mechanical meaning of the density matrix.

2. THE DENSITY MATRIX

Before entering the formal treatment of the density matrix (see Appendix B) let us build an intuitive picture. We begin with the simple system of two spin 1/2 nuclei, A and X, with its four energy levels E_1 to E_4 (Figure I.1) generally described in introductory NMR textbooks. We assume here (and throughout the book) a negative gyromagnetic ratio, γ. This explains the spin angular momentum orientation against the field in the lowest energy level E_4. Of course, in this state the magnetic moment is oriented with the field.

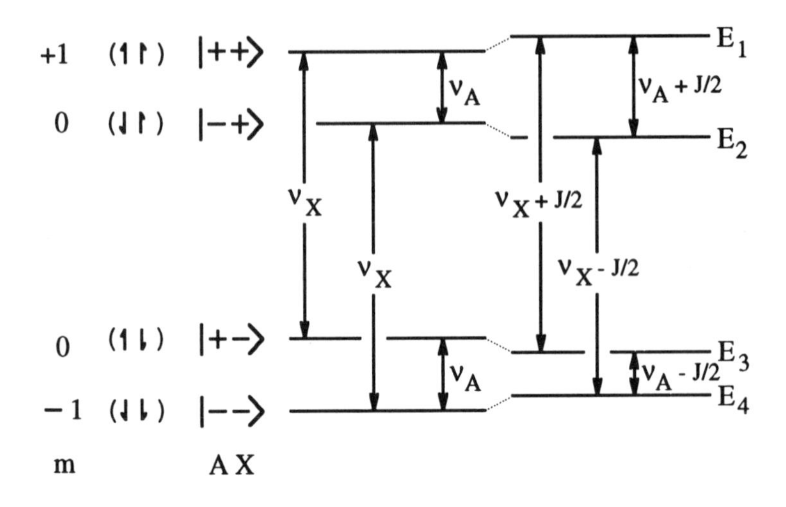

Figure I.1. Energy levels of an uncoupled (left) and coupled (right) heteronuclear AX system. The first column contains the total magnetic quantum number, m. Transition (precession) frequencies ν_A and ν_X and the coupling constant J are expressed in Hz.

The possible connections between the four quantum states represented by the "kets"

$$|++\rangle, |-+\rangle, |+-\rangle, |--\rangle,$$

are shown in Table I.1 (we assign the first symbol in the ket to nucleus A and the second, to nucleus X).

This is the general form of the density matrix for the system shown in Figure I.1. It can be seen that the off-diagonal elements of the matrix connect pairs of different states. These matrix elements are called "coherences" (for a formal definition see Appendix B) and are labeled according to the nature of the transitions between the corresponding states. For instance, in the transition

$$|++\rangle \rightarrow |-+\rangle$$

only the nucleus A is flipped. The corresponding matrix element will represent a *single quantum coherence* implying an A transition and will be labeled $1Q_A$. We thus find two $1Q_A$ and two $1Q_X$ coherences (the matrix elements on the other side of the diagonal do not represent other coherences; they are mirror images of the ones indicated above the diagonal). There is also one *double-quantum coherence*, $2Q_{AX}$, related to the transition

$$|++\rangle \rightarrow |--\rangle.$$

The *zero-quantum coherence* ZQ_{AX} can be considered as representing a flip-flop transition $E_2 \rightarrow E_3$. The name of this coherence does not necessarily imply that the energy of the transition is zero.

The diagonal elements represent populations.

Table I.1. Translation of the Classical Representation of a Two-spin System into a Density Matrix Representation

$\|AX\rangle$	$\|++\rangle$	$\|-+\rangle$	$\|+-\rangle$	$\|--\rangle$
$\|++\rangle$	P_1	$1Q_A$	$1Q_X$	$2Q_{AX}$
$\|-+\rangle$		P_2	ZQ_{AX}	$1Q_X$
$\|+-\rangle$			P_3	$1Q_A$
$\|--\rangle$				P_4

The density matrix contains complete information about the status of the ensemble of spins at a given time. Populations and macroscopic magnetizations can be derived from the elements of the density matrix, as we will see later. The reciprocal statement is not true: given the magnetization components and populations we do not have enough information to write all the elements of the density matrix. The extra information contained in the density matrix enables us to understand the NMR sequences which cannot be fully described by vector treatment.

3. THE DENSITY MATRIX DESCRIPTION OF A TWO-DIMENSIONAL HETERONUCLEAR CORRELATION SEQUENCE (2DHETCOR)

The purpose of 2DHETCOR is to reveal the pairwise correlation of different nuclear species (e.g., C-H or C-F) in a molecule. This is based on the scalar coupling interaction between the two spins.

3.1 Calculation Steps

Figure I.2 reveals that the density matrix treatment of a pulse sequence must include the following calculation steps:

- thermal equilibrium populations (off diagonal elements are zero)
- effects of rf pulses (rotation operators)
- evolution between pulses
- evolution during acquisition
- determination of observable magnetization.

Applying the sequence to an AX system (nucleus A is a ^{13}C, nucleus X is a proton) we will describe in detail each of these steps.

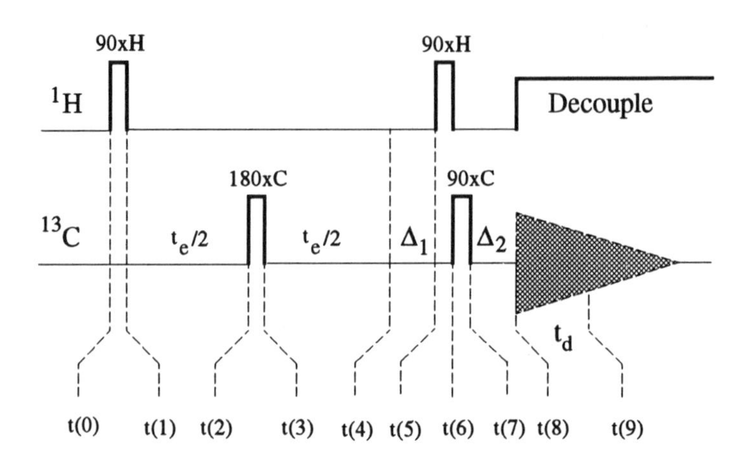

Figure I.2. The two-dimensional heteronuclear correlation sequence: $90xH - t_e/2 - 180xC - t_e/2 - \Delta_1 - 90xH - 90xC - \Delta_2 - \mathrm{AT}$.

3.2 Equilibrium Populations

At thermal equilibrium the four energy levels shown in Figure I.1 are populated according to the Boltzmann distribution law:

$$\frac{P_i}{P_j} = \frac{\exp(-E_i / kT)}{\exp(-E_j / kT)} = \exp\frac{E_j - E_i}{kT} \qquad (I.1)$$

Taking the least populated level as reference we have:

$$P_2 / P_1 = \exp[(E_1 - E_2) / kT] = \exp[h(\nu_A + J / 2) / kT] \qquad (I.2)$$

Since transition frequencies (10^8 Hz) are more than six orders of magnitude larger than coupling constants (tens or hundreds of Hz), we may neglect the latter (only when we calculate relative populations; of course, they will not be neglected when calculating transition frequencies). Furthermore, the ratios $h\nu_A/kT$ and $h\nu_X/kT$ are much smaller than 1. For instance, in a 4.7 Tesla magnet the ^{13}C Larmor frequency is $\nu_A = 50 \times 10^6$ Hz and

$$p = \frac{h\nu_A}{kT} = \frac{6.6 \cdot 10^{-34}\,\text{Js} \cdot 50 \cdot 10^6\,\text{s}^{-1}}{1.4 \cdot 10^{-23}(\text{J}/\text{K}) \cdot 300\text{K}} = 0.785 \cdot 10^{-5}$$

This justifies a first order series expansion [see (A11)]:

$$P_2 / P_1 = \exp(h\nu_A / kT) \cong 1 + (h\nu_A / kT) = 1 + p \qquad (I.3)$$

$$P_3 / P_1 = \exp(h\nu_X / kT) \cong 1 + (h\nu_X / kT) = 1 + q \qquad (I.4)$$

$$P_4 / P_1 \cong 1 + [h(\nu_A + \nu_X) / kT] = 1 + p + q \qquad (I.5)$$

In the particular case of the carbon-proton system the Larmor frequencies are in the ratio 1:4 (i.e., $q = 4p$).

We now normalize the sum of populations:

$$\begin{aligned} P_1 &= P_1 \\ P_2 &= (1+p)P_1 \\ P_3 &= (1+4p)P_1 \\ P_4 &= (1+5p)P_1 \\ \hline 1 &= P_1(4+10p) = P_1 S \end{aligned} \tag{I.6}$$

Hence,

$$\begin{aligned} P_1 &= 1/S \\ P_2 &= (1+p)/S \\ P_3 &= (1+4p)/S \\ P_4 &= (1+5p)/S \end{aligned}$$

where

$$S = 4 + 10p$$

Given the small value of p we can work with the approximation S≅4. Then the density matrix at equilibrium is:

$$D(0) = \begin{bmatrix} P_1 & 0 & 0 & 0 \\ 0 & P_2 & 0 & 0 \\ 0 & 0 & P_3 & 0 \\ 0 & 0 & 0 & P_4 \end{bmatrix} = \frac{1}{4}\begin{bmatrix} 1 & 0 & 0 & 0 \\ 0 & 1+p & 0 & 0 \\ 0 & 0 & 1+4p & 0 \\ 0 & 0 & 0 & 1+5p \end{bmatrix}$$

$$= \frac{1}{4}\begin{bmatrix} 1 & 0 & 0 & 0 \\ 0 & 1 & 0 & 0 \\ 0 & 0 & 1 & 0 \\ 0 & 0 & 0 & 1 \end{bmatrix} + \frac{p}{4}\begin{bmatrix} 0 & 0 & 0 & 0 \\ 0 & 1 & 0 & 0 \\ 0 & 0 & 4 & 0 \\ 0 & 0 & 0 & 5 \end{bmatrix}$$

It is seen that the first term of the sum above is very large compared to the second term. However, the first term is not important since it contains the unit matrix [see (A20)-(A21)] and is not affected by any evolution or rotation operator (see Appendix B). Though much smaller, it is the second term which counts because it contains the population differences (*Vive la difference!*). From now on we will work with this term only, ignoring the constant factor $p/4$ and taking the license to continue to call it $D(0)$:

$$D(0)=\begin{bmatrix}0&0&0&0\\0&1&0&0\\0&0&4&0\\0&0&0&5\end{bmatrix} \tag{I.7}$$

Equilibrium density matrices for systems other than C-H can be built in exactly the same way.

3.3 The First Pulse

At time $t(0)$ a 90° proton pulse is applied along the x-axis. We now want to calculate $D(1)$, the density matrix after the pulse. The standard formula for this operation,

$$D(1) = R^{-1}\, D(0)\, R, \tag{I.8}$$

is explained in Appendix B. The rotation operator, R, for this particular case is [see (C18)]:

$$R_{90xH}=\frac{1}{\sqrt{2}}\begin{bmatrix}1&0&i&0\\0&1&0&i\\i&0&1&0\\0&i&0&1\end{bmatrix} \tag{I.9}$$

where $i=\sqrt{-1}$ is the imaginary unit.

Its inverse (reciprocal), R^{-1}, is readily calculated by transposition and conjugation [see (A22)-(A23)]:

$$R^{-1}_{90xH}=\frac{1}{\sqrt{2}}\begin{bmatrix}1&0&-i&0\\0&1&0&-i\\-i&0&1&0\\0&-i&0&1\end{bmatrix} \tag{I.10}$$

First we multiply $D(0)$ by R. Since the matrix multiplication is not commutative (see Appendix A for matrix multiplication rules), it is necessary to specify that we *postmultiply* $D(0)$ by R:

$$D(0)R = \begin{bmatrix} 0 & 0 & 0 & 0 \\ 0 & 1 & 0 & 0 \\ 0 & 0 & 4 & 0 \\ 0 & 0 & 0 & 5 \end{bmatrix} \frac{1}{\sqrt{2}} \begin{bmatrix} 1 & 0 & i & 0 \\ 0 & 1 & 0 & i \\ i & 0 & 1 & 0 \\ 0 & i & 0 & 1 \end{bmatrix}$$

$$= \frac{1}{\sqrt{2}} \begin{bmatrix} 0 & 0 & 0 & 0 \\ 0 & 1 & 0 & i \\ 4i & 0 & 4 & 0 \\ 0 & 5i & 0 & 5 \end{bmatrix} \tag{I.11}$$

Then we *premultiply* the result by R^{-1}:

$$D(1) = R^{-1}[D(0)R] = \frac{1}{\sqrt{2}} \begin{bmatrix} 1 & 0 & -i & 0 \\ 0 & 1 & 0 & -i \\ -i & 0 & 1 & 0 \\ 0 & -i & 0 & 1 \end{bmatrix} \frac{1}{\sqrt{2}} \begin{bmatrix} 0 & 0 & 0 & 0 \\ 0 & 1 & 0 & i \\ 4i & 0 & 4 & 0 \\ 0 & 5i & 0 & 5 \end{bmatrix}$$

$$= \frac{1}{2} \begin{bmatrix} 4 & 0 & -4i & 0 \\ 0 & 6 & 0 & -4i \\ 4i & 0 & 4 & 0 \\ 0 & 4i & 0 & 6 \end{bmatrix} = \begin{bmatrix} 2 & 0 & -2i & 0 \\ 0 & 3 & 0 & -2i \\ 2i & 0 & 2 & 0 \\ 0 & 2i & 0 & 3 \end{bmatrix} \tag{I.12}$$

It is good to check this result by making sure that the matrix $D(1)$ is Hermitian, i.e., every matrix element below the main diagonal is the complex conjugate of its corresponding element above the diagonal [see (A24)] (neither the rotation operators, nor the partial results need be Hermitian). *Comparing $D(1)$ to $D(0)$ we see that the 90° proton pulse created proton single-quantum coherences, did not touch the carbon, and redistributed the populations.*

3.4 Evolution from t (1) to t (2)

The standard formula[1] describing the time evolution of the density matrix elements in the absence of a pulse is:

$$d_{mn}(t) = d_{mn}(0)\exp(-i\omega_{mn}t) \qquad \text{(I.13)}$$

d_{mn} is the matrix element (row m, column n) and $\omega_{mn}=(E_m - E_n)/\hbar$ is the angular frequency of the transition $m \rightarrow n$.

We observe that during evolution the diagonal elements are invariant since $\exp[i(E_m - E_m)/\hbar] = 1$. The off diagonal elements experience a periodic evolution. Note that $d_{mn}(0)$ is the starting point of the evolution immediately after a given pulse. In the present case, the elements $d_{mn}(0)$ are those of $D(1)$.

We now want to calculate $D(2)$ at the time $t(2)$ shown in Figure I.2. We have to consider the evolution of elements d_{13} and d_{24}. In a frame rotating with the proton transmitter frequency ω_{trH}, after an evolution time $t_e/2$, their values are:

$$d_{13} = -2i\exp(-i\Omega_{13}t_e / 2) = B \qquad \text{(I.14)}$$

$$d_{24} = -2i\exp(-i\Omega_{24}t_e / 2) = C \qquad \text{(I.15)}$$

where $\Omega_{13} = \omega_{13} - \omega_{trH}$ and $\Omega_{24} = \omega_{24} - \omega_{trH}$.

Hence

$$D(2) = \begin{bmatrix} 2 & 0 & B & 0 \\ 0 & 3 & 0 & C \\ B* & 0 & 2 & 0 \\ 0 & C* & 0 & 3 \end{bmatrix} \qquad \text{(I.16)}$$

$B*$ and $C*$ are the complex conjugates of B and C (see Appendix A).

---

[1]In our treatment, relaxation during the pulse sequence is ignored. This contributes to a significant simplification of the calculations without affecting the main features of the resulting 2D spectrum.

3.5 The Second Pulse

The rotation operators for this pulse are [see(C17)]:

$$R_{180xC} = \begin{bmatrix} 0 & i & 0 & 0 \\ i & 0 & 0 & 0 \\ 0 & 0 & 0 & i \\ 0 & 0 & i & 0 \end{bmatrix} \quad (I.17) \quad ; R_{180xC}^{-1} = \begin{bmatrix} 0 & -i & 0 & 0 \\ -i & 0 & 0 & 0 \\ 0 & 0 & 0 & -i \\ 0 & 0 & -i & 0 \end{bmatrix} \quad (I.18)$$

Postmultiplying $D(2)$ by R gives:

$$D(2)R_{180xC} = \begin{bmatrix} 0 & 2i & 0 & iB \\ 3i & 0 & iC & 0 \\ 0 & i & iB^* & 2i \\ iC^* & 0 & 3i & 0 \end{bmatrix} \quad (I.19)$$

Premultiplying (I.19) by R^{-1} gives:

$$D(3) = \begin{bmatrix} 3 & 0 & C & 0 \\ 0 & 2 & 0 & B \\ C^* & 0 & 3 & 0 \\ 0 & B^* & 0 & 2 \end{bmatrix} \quad (I.20)$$

Comparing $D(3)$ with $D(2)$ we note that the 180° pulse on carbon has caused a population inversion (interchange of d_{11} and d_{22}). It has also interchanged the coherences B and C (d_{13} and d_{24}). This means that B, after having evolved with the frequency ω_{13} during the first half of the evolution time [see (I.14)], *will now evolve with the frequency ω_{24}, while C switches form ω_{24} to ω_{13}.*

3.6 Evolution from $t\,(3)$ to $t\,(4)$

According to (I.13) the elements $d_{13}$and d_{42} become:

$$d_{13} = C\exp(-i\Omega_{13}t_e / 2) \qquad (I.21)$$

$$d_{24} = B\exp(-i\Omega_{24}t_e / 2) \qquad (I.22)$$

From Figure I.1 we see that in the laboratory frame

$$\omega_{13} = 2\pi(\nu_X + J/2) = \omega_H + \pi J \qquad (I.23)$$

$$\omega_{24} = 2\pi(\nu_X - J/2) = \omega_H - \pi J \qquad (I.24)$$

In the rotating frame (low case) ω becomes (capital) Ω. Taking the expressions of B and C from (I.14) and (I.15), relations (I.21) and (I.22) become

$$\begin{aligned} d_{13} &= -2i\exp[-i(\Omega_H - \pi J)t_e / 2]\exp[-i(\Omega_H + \pi J)t_e / 2] \\ &= -2i\exp(-i\Omega_H t_e) \end{aligned} \qquad (I.25)$$

$$d_{24} = -2i\exp(-i\Omega_H t_e) = d_{13} \qquad (I.26)$$

None of the matrix elements of $D(4)$ contains the coupling constant J. The result looks like that of a *decoupled evolution.* The averaged shift Ω_H (center frequency of the doublet) is expressed while the coupling is not. We know that the coupling J was actually present during the evolution, as documented by the intermediate results $D(2)$ and $D(3)$. We call the sequence $t_e/2$ - $180C$ - $t_e/2$ a *refocusing routine.* The protons which were fast (Ω_{13}) during the first $t_e/2$ are slow (Ω_{24}) during the second $t_e/2$ and vice versa (they change label).

3.7 The Role of Δ_1

In order to understand the role of the supplementary evolution Δ_1 we have to carry on the calculations without it, i.e., with d_{13}=d_{24}. We find out (see Appendix I) that the useful signal is canceled. To obtain maximum signal, d_{13} and d_{24} must be equal but of opposite signs. This is what the delay Δ_1 enables us to achieve.

Evolution during Δ_1 yields:

$$\begin{aligned} d_{13}(5) &= d_{13}(4)\exp(-i\Omega_{13}\Delta_1) \\ &= -2i\exp(-i\Omega_H t_e)\exp[-i(\Omega_H + \pi J)\Delta_1] \\ &= -2i\exp[-i\Omega_H(t_e + \Delta_1)]\exp(-i\pi J\Delta_1) \end{aligned} \tag{I.27}$$

$$d_{24}(5) = -2i\exp[-i\Omega_H(t_e + \Delta_1)]\exp(+i\pi J\Delta_1) \tag{I.28}$$

To achieve our goal we choose Δ_1=1/2J, which implies $\pi J\Delta_1$=π/2.

Using the expression [see (A16)]

$$\exp(\pm i\pi / 2) = \cos(\pi / 2) \pm i\sin(\pi / 2) = \pm i$$

$$\begin{aligned} \exp(-i\pi J\Delta_1) &= -i \\ \exp(+i\pi J\Delta_1) &= +i \end{aligned} \tag{I.29}$$

We now have

$$\begin{aligned} d_{13}(5) &= -2\exp[-i\Omega_H(t_e + \Delta_1)] \\ d_{24}(5) &= +2\exp[-i\Omega_H(t_e + \Delta_1)] \end{aligned} \tag{I.30}$$

For the following calculations it is convenient to use the notations

$$\begin{aligned} c &= \cos[\Omega_H(t_e + \Delta_1)] \\ s &= \sin[\Omega_H(t_e + \Delta_1)] \end{aligned} \tag{I.31}$$

which lead to

$$\begin{aligned} d_{13}(5) &= -2(c - is) \\ d_{24}(5) &= +2(c - is) \end{aligned} \tag{I.32}$$

At this point the density matrix is:

$$D(5) = \begin{bmatrix} 3 & 0 & -2(c-is) & 0 \\ 0 & 2 & 0 & 2(c-is) \\ -2(c+is) & 0 & 3 & 0 \\ 0 & 2(c+is) & 0 & 2 \end{bmatrix} \quad \text{(I.33)}$$

3.8 Third and Fourth Pulses

Although physically these pulses are applied separately, we may save some calculation effort by treating them as a single nonselective pulse.

The expressions of R_{90xC} and R_{90xH} are taken from Appendix C.

$$R_{90xCH} = R_{90xC}R_{90xH}$$

$$= \frac{1}{\sqrt{2}} \begin{bmatrix} 1 & i & 0 & 0 \\ i & 1 & 0 & 0 \\ 0 & 0 & 1 & i \\ 0 & 0 & i & 1 \end{bmatrix} \frac{1}{\sqrt{2}} \begin{bmatrix} 1 & 0 & i & 0 \\ 0 & 1 & 0 & i \\ i & 0 & 1 & 0 \\ 0 & i & 0 & 1 \end{bmatrix}$$

$$= \frac{1}{2} \begin{bmatrix} 1 & i & i & -1 \\ i & 1 & -1 & i \\ i & -1 & 1 & i \\ -1 & i & i & 1 \end{bmatrix} \quad \text{(I.34)}$$

The reciprocal of (I.34) is:

$$R_{90xCH}^{-1} = \frac{1}{2} \begin{bmatrix} 1 & -i & -i & -1 \\ -i & 1 & -1 & -i \\ -i & -1 & 1 & -i \\ -1 & -i & -i & 1 \end{bmatrix}$$

$$D(5) \cdot R_{90xCH}$$

$$= \frac{1}{2}\begin{bmatrix} 3-2i(c-is) & 3i+2(c-is) & 3i-2(c-is) & -3-2i(c-is) \\ 2i-2(c-is) & 2+2i(c-is) & -2+2i(c-is) & 2i+2(c-is) \\ 3i-2(c+is) & -3-2i(c+is) & 3-2i(c+is) & 3i+2(c+is) \\ -2+2i(c+is) & 2i+2(c+is) & 2i-2(c+is) & 2+2i(c+is) \end{bmatrix}$$

Premultiplying the last result by R_{90xCH}^{-1} gives

$$D(7) = \frac{1}{2}\begin{bmatrix} 5 & i-4is & 0 & -4ic \\ -i+4is & 5 & 4ic & 0 \\ 0 & -4ic & 5 & i+4is \\ 4ic & 0 & -i-4is & 5 \end{bmatrix} \qquad \text{(I.35)}$$

Comparing $D(7)$ with $D(5)$ we make two distinct observations. First, as expected, carbon coherences are created in d_{12} and d_{34} due to the $90xC$ pulse. Second, the proton information [$s = \sin\Omega_H(t_e + \Delta_1)$] has been transferred from d_{13} and d_{24} into the carbon coherences d_{12} and d_{34}, which are

$$d_{12} = \frac{i - 4is}{2}$$

$$d_{34} = \frac{i + 4is}{2}$$

This is an important point of the sequence because now the *mixed* carbon and proton information can be carried into the final FID.

3.9 The Role of Δ_2

As we will see soon, the observable signal is proportional to the sum $d_{12} + d_{34}$. If we started the decoupled acquisition right at $t(7)$, the terms containing s would be cancelled. To save them, we allow for one more short coupled evolution Δ_2. Since no r.f. pulse follows after $t(7)$, we know that every matrix element will evolve in its own box according to (I.13). It is therefore sufficient, from now on, to follow the evolution of the carbon coherences d_{12} and d_{34} which constitute the observables in this sequence.

According to (I.13), at $t(8)$ coherences d_{12} and d_{34} become

$$d_{12}(8) = i(1/2 - 2s)\exp(-i\Omega_{12}\Delta_2) \quad \text{(I.36)}$$

$$d_{34}(8) = i(1/2 + 2s)\exp(-i\Omega_{34}\Delta_2) \quad \text{(I.37)}$$

where $\Omega_{12} = \omega_{12} - \omega_{trC}$ and $\Omega_{34} = \omega_{34} - \omega_{trC}$ indicate that now we are in the carbon rotating frame, which is necessary to describe the carbon signal during the free induction decay.

As shown in Figure I.1 the transition frequencies of carbon (nucleus A) are:

$$\nu_{12} = \nu_A + \frac{J}{2} = \nu_C + \frac{J}{2}$$

$$\nu_{34} = \nu_A - \frac{J}{2} = \nu_C - \frac{J}{2}$$

Since $\omega = 2\pi\nu$, and we are in rotating coordinates we obtain:

$$\Omega_{12} = \Omega_C + \pi J \quad \text{(I.38)}$$

$$\Omega_{34} = \Omega_C - \pi J \quad \text{(I.39)}$$

Hence,

$$d_{12}(8) = i(1/2 - 2s)\exp(-i\Omega_C\Delta_2)\exp(-i\pi J\Delta_2) \quad \text{(I.40)}$$

$$d_{34}(8) = i(1/2 + 2s)\exp(-i\Omega_C\Delta_2)\exp(+i\pi J\Delta_2) \quad \text{(I.41)}$$

Analyzing the role of Δ_2 in (I.40 – 41) we see that for $\Delta_2 = 0$ the terms in s which contain the proton information are lost when we calculate the sum of d_{12} and d_{34} As discussed previously for Δ_1, here also, the desired signal is best obtained for $\Delta_2 = 1/2J$, which leads to $\exp(\pm i\pi J\Delta_2) = \pm i$ and

$$d_{12}(8) = +(1/2 - 2s)\exp(-i\Omega_C\Delta_2) \quad \text{(I.42)}$$

$$d_{34}(8) = -(1/2 + 2s)\exp(-i\Omega_C\Delta_2) \quad \text{(I.43)}$$

3.10 Detection

From the time $t(8)$, on the system is proton decoupled, i.e., both d_{12} and d_{34} evolve with the frequency Ω_C:

$$d_{12}(9) = +(1/2 - 2s)\exp(-i\Omega_C\Delta_2)\exp(-i\Omega_C t_d) \qquad (I.44)$$

$$d_{34}(9) = -(1/2 + 2s)\exp(-i\Omega_C\Delta_2)\exp(-i\Omega_C t_d) \qquad (I.45)$$

Our density matrix calculations, carried out for every step of the sequence, have brought us to the relations (I.44-45). Now it is time to derive the observable (transverse) carbon magnetization components. This is done by using the relations (B19) and (B20) in Appendix B:

$$M_{TC} = M_{xC} + iM_{yC} = -(4M_{oC} / p)(d_{12}^* + d_{34}^*) \qquad (I.46)$$

The transverse magnetization M_T is a complex quantity which combines the x and y components of the magnetization vector. We must now reintroduce the factor $p/4$ which we omitted, for convenience, starting with (I.7). This allows us to rewrite (I.46) into a simpler form:

$$M_{TC} = -M_{oC}(d_{12}^* + d_{34}^*) \qquad (I.47)$$

By inserting (I.44-45) into (I.47) we obtain

$$M_{TC} = 4M_{oC}s\exp(i\Omega_C\Delta_2)\exp(i\Omega_C t_d) \qquad (I.48)$$

With the explicit expression of s (I.31):

$$M_{TC} = 4M_{oC}\sin[\Omega_H(t_e + \Delta_1)]\exp(i\Omega_C\Delta_2)\exp(i\Omega_C t_d) \qquad (I.49)$$

Equation (I.49) represents the final result of our 2DHETCOR analysis by means of the density matrix formalism and it contains all the information we need.

We learn from (I.49) *that the carbon magnetization rotates by* $\Omega_C t_d$ *while being amplitude modulated by the proton evolution* $\Omega_H t_e$. *Fourier transformation with respect to both time domains will yield the two-dimensional spectrum.*

The signal is enhanced by a factor of four, representing the γ_H/γ_C ratio. The polarization transfer achieved in 2DHETCOR *and other heteronuclear pulse sequences cannot be explained by the vector representation.*

When transforming with respect to t_d, all factors other than $\exp(i\Omega_C t_d)$ are regarded as constant. A single peak frequency, Ω_C, is obtained. When transforming with respect to t_e, all factors other than $\sin[\Omega_H(t_e + \Delta_1)]$ are regarded as constant. Since

$$\sin\alpha = \frac{e^{i\alpha} - e^{-i\alpha}}{2i} \tag{I.50}$$

both $+\Omega_H$ and $-\Omega_H$ are obtained (Figure I.3a).

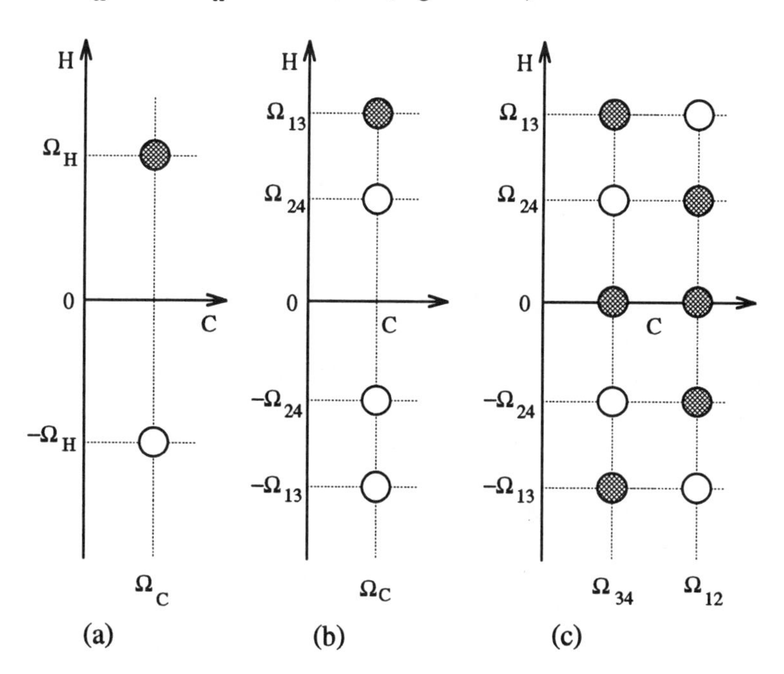

Figure I.3. Schematic 2D heteronuclear correlation spectra (contour plot): (a) fully decoupled, (b) proton decoupled during the acquisition, and (c) fully coupled. Filled and open circles represent positive and negative peaks. With the usually employed magnitude calculation (absolute value), all peaks are positive. See experimental spectra in figures 3.11, 3.9 and 3.7 of the book by Martin and Zektzer (see Suggested Readings).

Imagine now that in the sequence shown in Figure I.2 we did not apply the 180° pulse on carbon and suppressed Δ_1. During the evolution time t_e the proton is coupled to carbon. During the acquisition, the carbon is decoupled from proton. The result (see Appendix I) is that along the carbon axis we see a single peak, while along the proton axis we see a doublet due to the proton-carbon coupling. If we calculate the magnetization following the procedure shown before, we find:

$$M_{TC} = -2M_{oC}(\cos\Omega_{13}t_e - \cos\Omega_{13}t_e)\exp[i\Omega_C(t_d + \Delta_2)] \qquad \text{(I.51)}$$

Reasoning as for (I.49) we can explain the spectrum shown in Figure I.3b.

Finally, if we also suppress the decoupling during the acquisition and the delay Δ_2 we obtain (see Appendix I)

$$\begin{aligned} M_{TC} = &-iM_{oC}(1/2 - \cos\Omega_{13}t_e + \cos\Omega_{24}t_e)\exp(i\Omega_{12}t_d) \\ &-iM_{oC}(1/2 + \cos\Omega_{13}t_e - \cos\Omega_{24}t_e)\exp(i\Omega_{34}t_d) \end{aligned} \qquad \text{(I.52)}$$

which yields the spectrum shown in Figure I.3c.

The lower part of the spectra is not displayed by the instrument, but proper care must be taken to place the proton transmitter beyond the proton spectrum. Such a requirement is not imposed on the carbon transmitter, provided quadrature phase detection is used.

The peaks in the lower part of the contour plot (negative proton frequencies) can also be eliminated if a more sophisticated pulse sequence is used, involving phase cycling. If such a pulse sequence is used, the proton transmitter can be positioned at mid-spectrum as well. An example of achieving quadrature detection in the domain t_e is given in Section 6 (COSY with phase cycling).

So far we have treated the AX (CH) system. In reality, the proton may be coupled to one or several other protons. In the sequence shown in Figure I.2 there is no proton-proton decoupling. The 2D spectrum will therefore exhibit single resonances along the carbon axis, but multiplets corresponding to proton-proton coupling, along the proton axis. An example is given in Figure I.4a which represents the high field region of the 2DHECTOR spectrum of a molecule

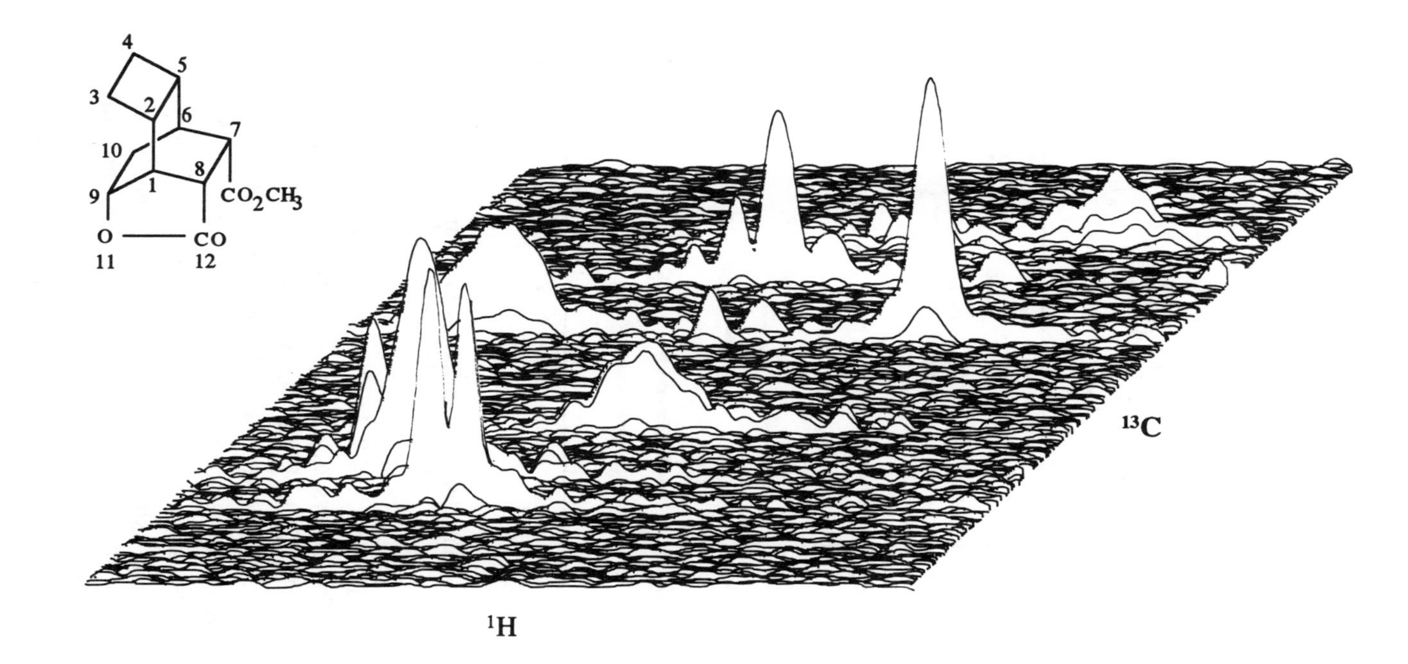

Figure I.4a. The high field region of the 2DHETCOR stack plot of a Nenitzescu's hydrocarbon derivative in $CDCl_3$ (at 50 MHz for ^{13}C). The peaks corresponding to the carbonyls, the methyl group, and the carbon in position 9 are at lower fields (M. Avram, G.D. Mateescu and C.D. Nenitzescu, unpublished work).

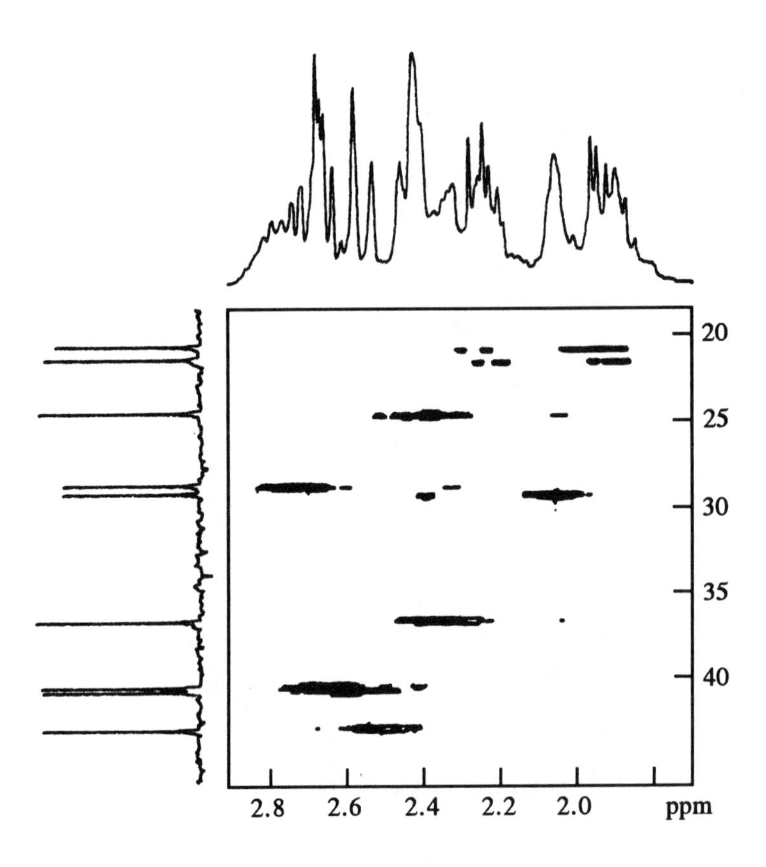

Figure I.4b. Contour plot of the spectrum in Figure I.4a.

formally derived from [4,2,2,$0^{2,5}$]deca-3,7,9-triene (Nenitzescu's hydrocarbon). The delays Δ_1 and Δ_2 were set to 3.6 ms in order to optimize the signals due to ^1J ($\cong$140 Hz). It should be noted that the relation (I.49) has been derived with the assumption that $\Delta_1 = \Delta_2 = \Delta = 1/2J$. For any other values of J the signal intensity is proportional to $\sin^2\pi J\Delta$. Thus, signals coming from long range couplings will have very small intensities. Figure I.4b is a contour plot of the spectrum shown in Figure I.4a. It shows in a more dramatic manner the advantage of 2D spectroscopy: the carbon-proton correlation and the disentangling of the heavily overlapping proton signals.

3.11 Comparison of the DM Treatment with Vector Representation

It is now possible to follow the 2DHETCOR vector representation (Figures I.6a through I.6d) and identify each step with the corresponding density matrix. It will be seen that one cannot draw the vectors for the entire sequence without the knowledge of the DM results.

As demonstrated in Appendix B (see B15-B22) the magnetization components at any time are given by:

$$M_{zA} = -(2M_{oA} / p)(d_{11} - d_{22} + d_{33} - d_{44}) \qquad \text{(B15)}$$

$$M_{zX} = -(2M_{oX} / q)(d_{11} - d_{33} + d_{22} - d_{44}) \qquad \text{(B21)}$$

$$M_{TA} = -(4M_{oA} / p)(d_{12}^{*} + d_{34}^{*}) \qquad \text{(B20)}$$

$$M_{TX} = -(4M_{oX} / q)(d_{13}^{*} + d_{24}^{*}) \qquad \text{(B22)}$$

Considering the simplification we made in (I.7), we must multiply the expressions above with the factor $p/4$. Also, remembering that for the CH system $q = 4p$, we obtain

$$M_{zC} = -(M_{oC} / 2)(d_{11} - d_{22} + d_{33} - d_{44})$$

$$M_{zH} = -(M_{oH} / 8)(d_{11} - d_{33} + d_{22} - d_{44})$$

$$M_{TC} = -M_{oC}(d_{12}^{*} + d_{34}^{*}) \qquad \text{(I.53)}$$

$$M_{TH} = -(M_{oH} / 4)(d_{13}^{*} + d_{24}^{*})$$

We will use the relations (I.53) throughout the sequence in order to find the magnetization components from the matrix elements.

At time $t(0)$ the net magnetization is in the z-direction for both proton and carbon. Indeed, with the matrix elements of $D(0)$ (see I.7) we find

$$M_{zC} = -(M_{oC} / 2)(0 - 1 + 4 - 5) = +M_{oC}$$

$$M_{zH} = -(M_{oH} / 8)(0 - 4 + 1 - 5) = +M_{oH} \qquad \text{(I.54)}$$

The transverse magnetizations M_{TC} and M_{TH} are both zero (all off-diagonal elements are zero), consistent with the fact that no pulse has been applied.

Although they are indiscernible at thermal equilibrium, we will now define *fast* and *slow* components using Figure I.5.

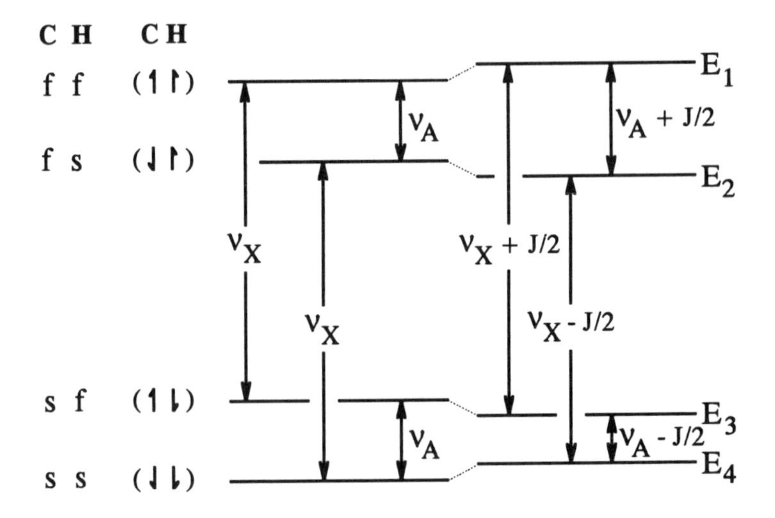

Figure I.5. *Fast* **and** *slow* **labeling.**

It is seen that protons in states 1 and 3 cannot be involved but in the higher frequency transition 1 – 3, i.e., they are *fast.* Those in states 2 and 4 are *slow.* Likewise, carbons in states 1 and 2 are *fast,* those in states 3 and 4 are *slow.* Therefore, according to (I.53) and (I.54), at $t(0)$ half of M_{zH} is due to *fast* protons ($d_{11} - d_{33}$) and the other half to *slow* protons. The *fast* and *slow* components of the proton magnetization are marked in Figure I.6 with 13 and 24, respectively. For carbon, it is 12 and 34.

Speaking of C-H pairs, a proton can add or subtract to the field "seen" by the carbon. Therefore the carbon will be *fast* if it pairs with a spin-up proton or *slow* if it pairs with a spin-down proton. The carbon spins will have a similar effect on protons. Figure I.5 shows that the spins become *faster* or *slower* by $J/2$ Hz.

Immediately after the 90*xH* pulse proton coherences were created. Using the matrix elements of *D*(1) (see I.12) we obtain:

$$M_{TH} = -(M_{oH} / 4)(2i + 2i) = -iM_{oH}$$

This tells us that the pulse brought the proton magnetization on the $-y$-axis (the reader is reminded that in the transverse magnetization, $M_T = M_x + iM_y$, the real part represents vectors along the x-axis and the imaginary part vectors along the y-axis). It can be verified that the longitudinal proton magnetization is zero since $d_{11} - d_{22} + d_{33} - d_{44} = 2 - 2 + 3 - 3 = 0$ (cf I.53). The carbon magnetization was not affected [Figure I.6a $t(1)$].

The chemical shift evolution $\Omega_H t_e/2$ is the average of the fast and slow evolutions discussed above [see (I.23 − 24)]. The vector 13 is ahead by $+\pi J t_e/2$, while 24 is lagging by the same angle (i.e., $-\pi J t_e/2$). The DM results [see (I.14 − 16)] demonstrate the same thing:

$$\begin{aligned} M_{TH} &= -(M_{oH} / 4)(B^* + C^*) \\ &= -(M_{oH} / 4)[2i \exp(i\Omega_{13} t_e / 2) + 2i \exp(i\Omega_{24} t_e / 2)] \\ &= -i(M_{oH} / 4) \exp(i\Omega_H t_e / 2)[\exp(i\pi J t_e / 2) + \exp(-i\pi J t_e / 2)] \end{aligned}$$

The carbon is still not affected [Figure I.6a $t(2)$].

Figure I.6b $t(3)$ tells us that the 180*xC* pulse reverses the carbon magnetization and also reverses the proton labels. As discussed above, the protons coupled to *up* carbons are fast and those coupled to *down* carbons are slow. Therefore inverting carbon orientation results in changing fast protons into slow protons and vice versa. This is mathematically documented in the DM treatment [see (I.20)]. The matrix element *B* is transferred in the slow (24) "slot" and will evolve from now on with the *slow* frequency Ω_{24}. The reverse is happening to the matrix element *C*. The longitudinal carbon magnetization changed sign:

$$-(d_{11} - d_{22} + d_{33} - d_{44}) = -(3 - 2 + 3 - 2) = -2$$

Figure I.6b $t(4)$ clearly shows that the second evolution $t_e/2$ completes the decoupling of proton from carbon. The fast vector 13 catches up with the slow 24 and at $t(4)$ they coincide. They have both precessed a total angle $\Omega_H t_e$ from their starting position along $-y$. We can verify that the matrix elements d^*_{13} and d^*_{24} are equal at $t(4)$ [see (I.25) and (I.26)]. The transverse magnetization, calculated from the matrix elements, is

$$M_{TH} = -(M_{oH}/4)[2i\exp(i\Omega_H t_e) + 2i\exp(i\Omega_H t_e)] = -iM_{oH}\exp(i\Omega_H t_e)$$

After separating the real and imaginary parts in M_{TH} we obtain

$$M_{TH} = -iM_{oH}(\cos\Omega_H t_e + i\sin\Omega_H t_e) = M_{oH}(\sin\Omega_H t_e - i\cos\Omega_H t_e)$$
$$M_{xH} = \text{real part of } M_{TH} = M_{oH}\sin\Omega_H t_e$$
$$M_{yH} = \text{coefficient of the imaginary part of } M_{TH} = -M_{oH}\cos\Omega_H t_e$$

This is in full accordance with the vector representation. The carbon magnetization is still along $-z$.

Figure I.6b $t(5)$ shows what happened during the delay Δ_1 which has been chosen equal to $1/2J$. Each of the two proton magnetization components rotated by an angle $\Omega_H \Delta_1$ but, with respect to the average, the fast component has gained $\pi/2$ while the slow one has lost $\pi/2$. As a result, the vectors are now opposite. We can verify [see (I.30)] that at $t(5)$ the elements d_{13} and d_{24} are equal and of opposite signs. The carbon magnetization did not change.

Figure I.6c shows the situation at $t(6)$, after the $90xH$ pulse. When we went through the DM treatment, we combined the last two pulses into a single rotation operator and this brought us directly from $D(5)$ to $D(7)$. However, for the comparison with the vector representation and for an understanding of the polarization transfer, it is necessary to discuss the density matrix $D(6)$. It can be calculated by applying the rotation operator R_{90xH} [see (I.9)] to $D(5)$ given in (I.33). The result is

$$D(6) = \begin{bmatrix} 3-2s & 0 & -2c & 0 \\ 0 & 2+2s & 0 & 2c \\ -2c & 0 & 3+2s & 0 \\ 0 & 2c & 0 & 2-2s \end{bmatrix} \tag{I.55}$$

The magnetization components in Figure I.6c are derived from the matrix elements of $D(6)$, using (I.53). What happens to the proton magnetization can be predicted from the previous vector representation but *what happens to the carbon cannot.* As far as the proton is concerned, its x components are not affected, while the other components rotate from y to z and from $-y$ to $-z$, as expected after a $90x$ pulse.

The net longitudinal carbon magnetization does not change, it is still $-M_{oC}$, but a sizable imbalance is created between its fast and slow components:

$$M_{12} = -(M_{oC}/2)(d_{11} - d_{22}) = -(M_{oC}/2)(3 - 2s - 2 - 2s)$$

$$= M_{oC}(-1/2 + 2s) \tag{I.56}$$

$$M_{34} = -(M_{oC}/2)(d_{33} - d_{44}) = -(M_{oC}/2)(3 + 2s - 2 + 2s)$$

$$= M_{oC}(-1/2 - 2s) \tag{I.57}$$

$$M_{zC} = M_{12} + M_{34} = -M_{oC} \tag{I.58}$$

The imbalance term is proportional to $s = \sin[\Omega_H(t_e + \Delta_1)]$, i.e, it is proton modulated. When s varies from +1 to -1, the quantity $2sM_{oC}$ varies from $+2M_{oC}$ to $-2M_{oC}$, a swing of $4M_{oC}$. The remaining of the sequence is designed to make this modulated term observable.

For s greater than 1/4, M_{12} becomes positive while M_{34} remains negative. In other words, the fast carbons are now predominantly up and the slow ones predominantly down. A correlation has been created between the *up-down* and the *fast-slow* quality of the carbon spins. The Figure I.6c is drawn for $s \cong 0.95$

The last pulse of the sequence, a $90xC$, brings us to $D(7)$ [see (I.35)] and to Figure I.6d $t(7)$. It is seen that the vector representation can explain how carbon magnetization is affected by the pulse (12 goes in $-y$ and 34 in $+y$), but *it cannot explain the nulling of proton magnetization.* Note that from $t(5)$ on, the net magnetization was zero (opposite vectors) but now 13 and 24 are null themselves. The density matrix $D(7)$ [see(I.35)] shows

$$M_{T12} = iM_{oC}(1/2 - 2s)$$

$$M_{T34} = iM_{oC}(1/2 + 2s)$$

Separation of the real and imaginary parts gives

$$M_{x12} = 0 \quad ; \quad M_{y12} = M_{oC}(1/2 - 2s)$$
$$M_{x34} = 0 \quad ; \quad M_{y34} = M_{oC}(1/2 + 2s)$$

The net carbon magnetization is

$$M_{xC} = M_{x12} + M_{x34} = 0$$
$$M_{yC} = M_{y12} + M_{y34} = M_{oC}$$

The proton modulated term, $2s$, is not yet observable (it does not appear in M_{xC} or M_{yC}). The delay Δ_2 will render it observable. We see in Figure I.6d $t(8)$ that the two components have rotated by an average of $\Omega_C\Delta_2$. The fast one has gained $\pi/2$ and the slow one has lost $\pi/2$. As a result, the two vectors (of unequal magnitude) are now coincident. Relations (I.42) and (I.43) confirm that at time $t(8)$ both matrix elements d^*_{12} and d^*_{34} have the same phase factor, $\exp(i\Omega_C\Delta_2)$. The net carbon magnetization is

$$M_{TC} = -M_{oTC}(d^*_{12} + d^*_{34}) = -M_{oC}(1/2 - 2s - 1/2 - 2s)\exp(i\Omega_C\Delta_2)$$
$$= 4sM_{oC}\exp(i\Omega_C\Delta_2) \qquad \text{(I.59)}$$

The factor 4 in (I.59) represents the enhancement of the carbon magnetization by polarization transfer. This could not be even guessed from the vector representation.

Nothing remarkable happens after the end of Δ_2. The proton decoupler is turned on and the carbon magnetization is precessing as a whole during the detection time t_d (no spreadout of the fast and slow components).

This is the end of the vector representation of the 2DHETCOR sequence. Such representation would not have been possible without the *complete* information provided by the DM treatment.

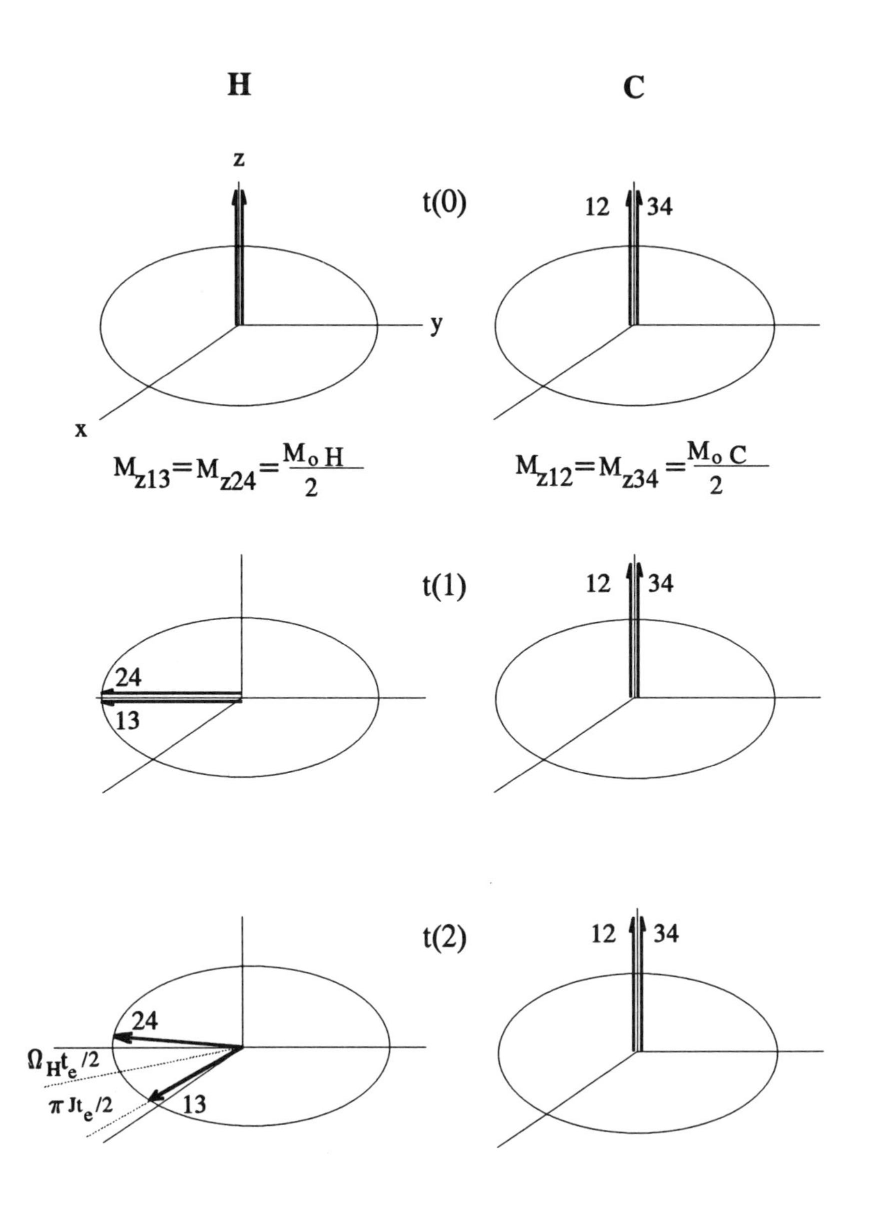

Figure I.6a. Vector representation of 2DHETCOR from *t*(0) to *t*(2). The magnetization vectors are arbitrarily taken equal for C and H in order to simplify the drawing. Actually the ^{1}H magnetization at equilibrium is 16 times larger than that of ^{13}C.

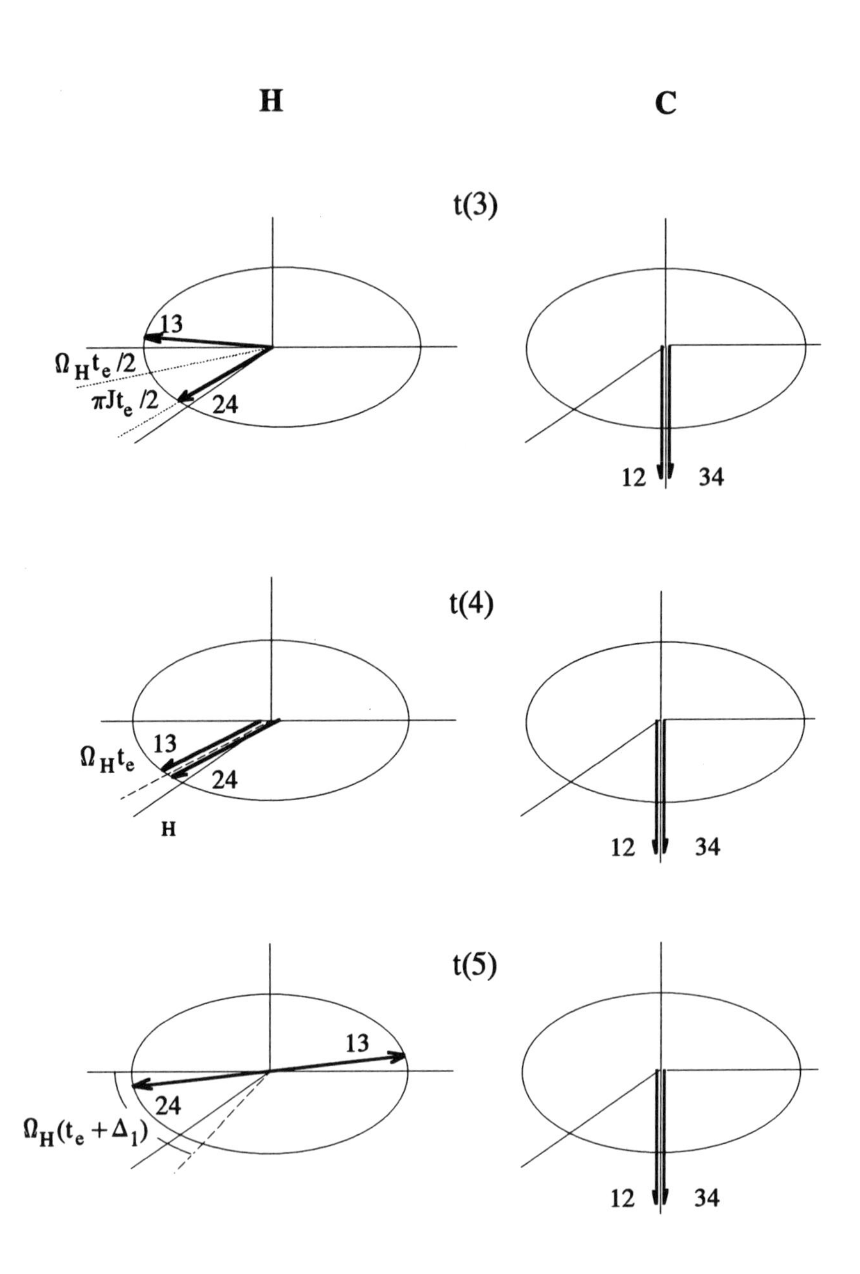

Figure I.6b. Vector representation of 2DHETCOR from *t*(3) to *t*(5).

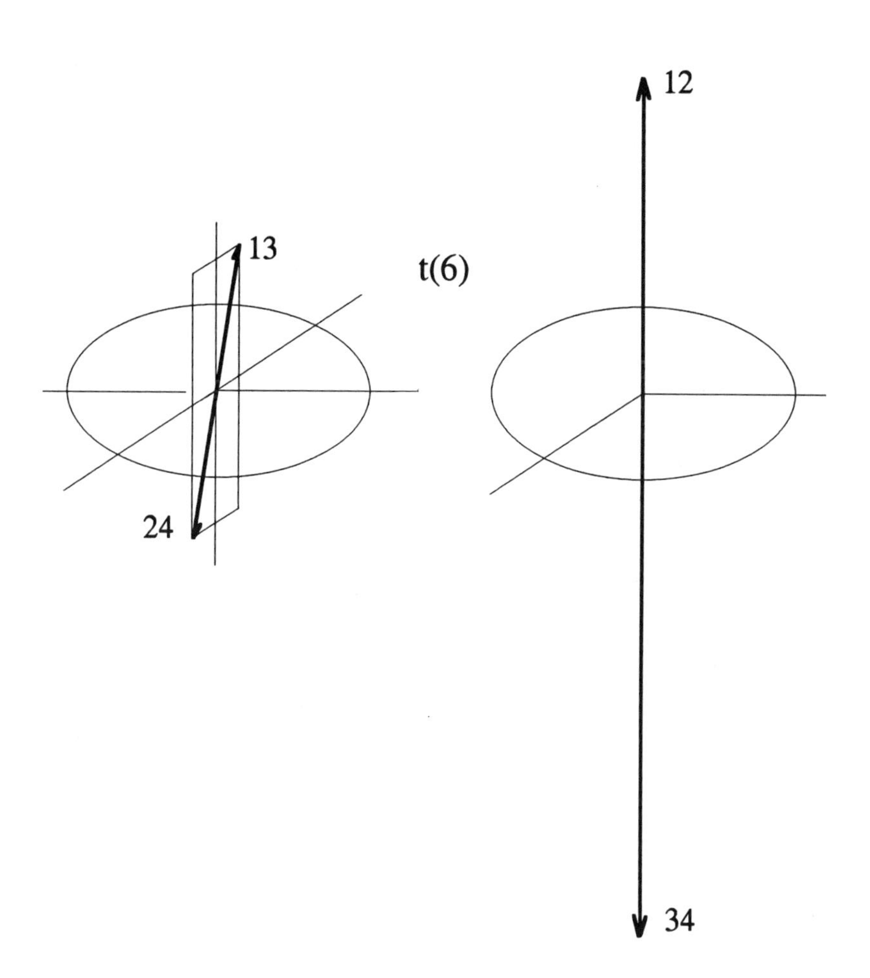

Figure I.6c. Vector representation of 2DHETCOR at time $t(6)$. The carbon magnetization components depend on the value of s (they are proton modulated):

$$M_{z12} = -(1 - 4s) M_{oC} / 2$$

$$M_{z34} = -(1 + 4s) M_{oC} / 2$$

$$s = \sin \Omega_H (t_e + \Delta_1)$$

The figure is drawn for $s \cong 0.95$.

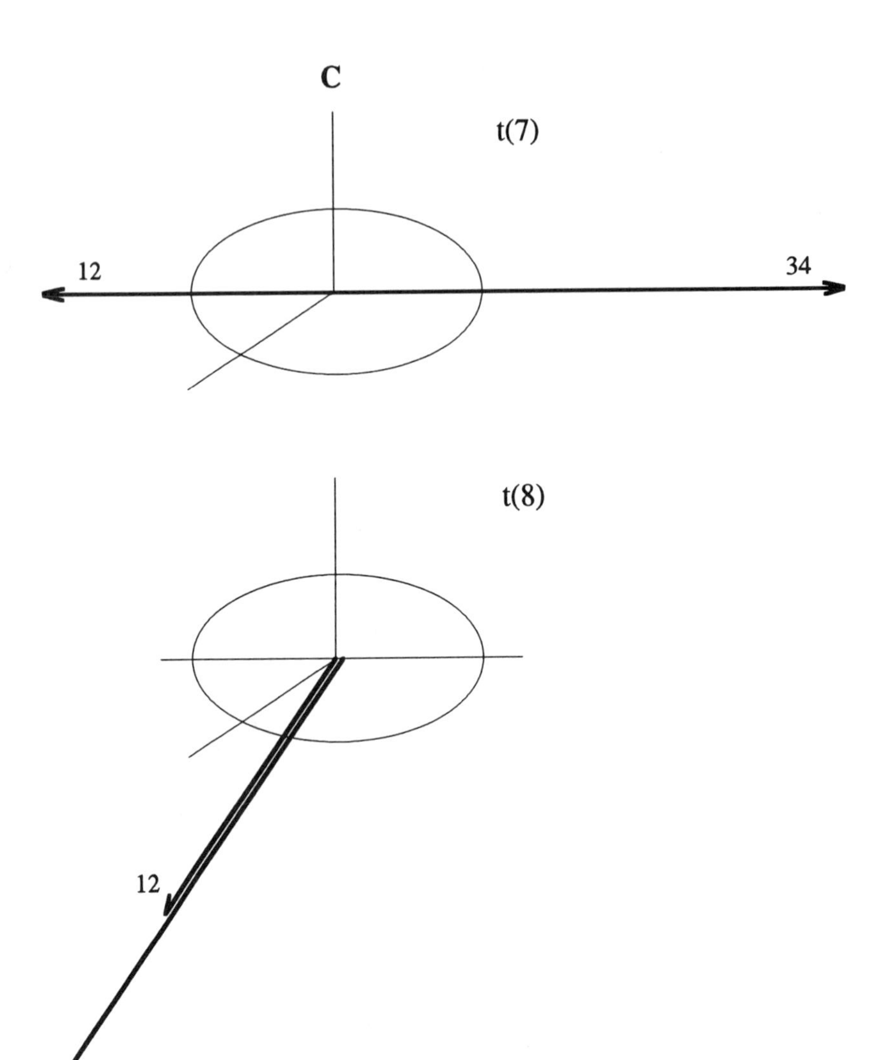

Figure I.6d. Vector representation of 2DHETCOR at $t(7)$ and $t(8)$. The carbon magnetization components at $t(7)$ are

$$M_{y12} = (1 - 4s) M_{oC} / 2$$

$$M_{y34} = (1 + 4s) M_{oC} / 2$$

$$s = \sin \Omega_H (t_e + \Delta_1)$$

The proton magnetization components, both fast and slow, have vanished.

4. THE DENSITY MATRIX DESCRIPTION OF A DOUBLE-QUANTUM COHERENCE EXPERIMENT (INADEQUATE)

The main goal of INADEQUATE (Incredible Natural Abundance DoublE QUAntum Transfer Experiment) is to eliminate the strong signal of noncoupled ^{13}C nuclei in order to easily observe the 200 times weaker satellites due to C-C coupling. This is realized by exploiting the different phase responses of the coupled and non-coupled spin signals when the phase of the observe pulse is varied (see Figure I.7). The receiver phase is matched with the desired signal. It shall be seen that the different phase behavior of the coupled nuclei is connected with their double-quantum coherence. The beauty of INADEQUATE resides in its basic simplicity: only a two-step cycle is theoretically needed to eliminate the unwanted signal. That the real life sequences may reach 128 or more steps is exclusively due to hardware (pulse) imperfections whose effects must be corrected by additional phase cycling.

The essence of INADEQUATE can be understood by following the basic sequence shown in Figure I.7.

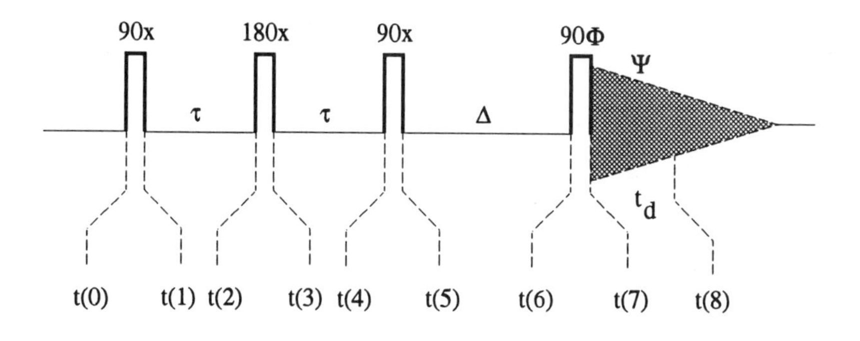

Figure I.7. The INADEQUATE sequence: $90x - \tau - 180x - \tau - 90x - \Delta - 90\Phi - AT$ (proton decoupling is applied throughout the experiment).

4.1 Equilibrium Populations

At thermal equilibrium the four energy levels shown in Figure I.8 are populated according to the Boltzmann distribution law, as shown in (I.1) through (I.5). In this case both ν_A and ν_X are ^{13}C transition frequencies. The difference between ν_A and ν_X, due to different chemical shifts, is too small to be taken into account when calculating the populations. We assume $q = p$ and (I.6) becomes:

$$\begin{aligned} P_1 &= P_1 \\ P_2 &= (1+p)P_1 \\ P_3 &= (1+p)P_1 \\ P_4 &= (1+2p) / P_1 \end{aligned}$$

$$1 = (4+4p)P_1 = P_1 S$$

Hence,

$$\begin{aligned} P_1 &= 1/S \\ P_2 = P_3 &= (1+p)/S \\ P_4 &= (1+2p)/S \end{aligned}$$

where

$$S = 4 + 4p \cong 4$$

and the density matrix at equilibrium is:

$$D(0) = \begin{bmatrix} P_1 & 0 & 0 & 0 \\ 0 & P_2 & 0 & 0 \\ 0 & 0 & P_3 & 0 \\ 0 & 0 & 0 & P_4 \end{bmatrix} = \frac{1}{4}\begin{bmatrix} 1 & 0 & 0 & 0 \\ 0 & 1+p & 0 & 0 \\ 0 & 0 & 1+p & 0 \\ 0 & 0 & 0 & 1+2p \end{bmatrix}$$

$$= \frac{1}{4}\begin{bmatrix} 1 & 0 & 0 & 0 \\ 0 & 1 & 0 & 0 \\ 0 & 0 & 1 & 0 \\ 0 & 0 & 0 & 1 \end{bmatrix} + \frac{p}{4}\begin{bmatrix} 0 & 0 & 0 & 0 \\ 0 & 1 & 0 & 0 \\ 0 & 0 & 1 & 0 \\ 0 & 0 & 0 & 2 \end{bmatrix} \qquad (I.60)$$

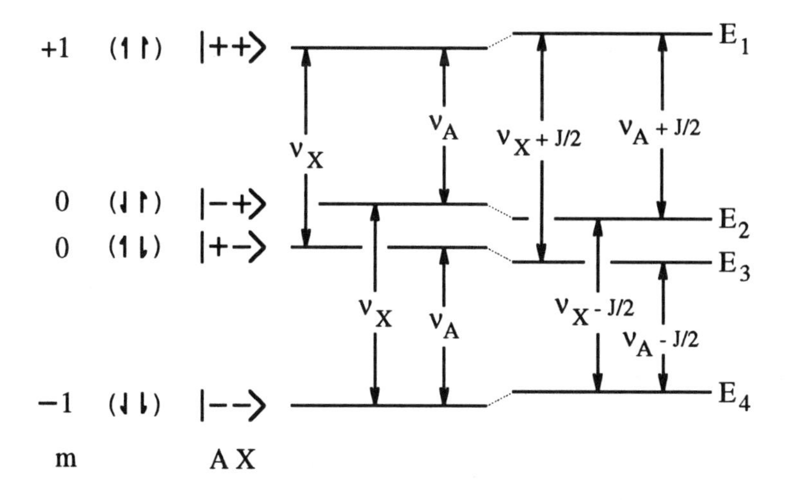

Figure I.8. Energy levels of a homonuclear AX system (noncoupled and coupled). Transition frequencies and coupling constants are in Hz.

We will again ignore the (large) first term which is not affected by pulses or evolution, put aside the constant factor $p/4$ and start with

$$D(0) = \begin{bmatrix} 0 & 0 & 0 & 0 \\ 0 & 1 & 0 & 0 \\ 0 & 0 & 1 & 0 \\ 0 & 0 & 0 & 2 \end{bmatrix} \qquad (I.61)$$

In order to compare the results of the density matrix treatment with those of the vectorial representation, we will calculate for every step of the sequence the magnetization components, using the relations (B15) – (B22). We must also consider that here $q = p$ and that $M_{oA} = M_{oX} = M_o/2$, where M_o refers to magnetization due to adjacent ^{13}C atoms A and X. Thus our magnetization equations become (cf. I.53):

$$\begin{aligned} M_{zA} &= -(M_o/4)(d_{11} - d_{22} + d_{33} - d_{44}) \\ M_{zX} &= -(M_o/4)(d_{11} - d_{33} + d_{22} - d_{44}) \\ M_{TA} &= -(M_o/2)(d_{12}^* + d_{34}^*) \\ M_{TX} &= -(M_o/2)(d_{13}^* + d_{24}^*) \end{aligned} \qquad (I.62)$$

One can check that at thermal equilibrium, when $D = D(0)$

$$M_{zA} = -(M_o / 4)(0-1+1-2) = M_o / 2$$
$$M_{zX} = -(M_o / 4)(0-1+1-2) = M_o / 2$$

The transverse magnetization

$$M_{TA} = M_{TX} = 0$$

4.2 The First Pulse

At time $t(0)$ a nonselective pulse $90xAX$ is applied. Since all pulses in this sequence are nonselective, the notation AX will be omitted. The density matrix $D(1)$ after the pulse is calculated according to:

$$D(0) = R^{-1}D(0)R$$

The rotation operator R and its reciprocal R^{-1} for the nonselective $90x$ pulse have been calculated in (I.34):

$$R = \frac{1}{2}\begin{bmatrix} 1 & i & i & -1 \\ i & 1 & -1 & i \\ i & -1 & 1 & i \\ -1 & i & i & 1 \end{bmatrix} \quad ; \quad R^{-1} = \frac{1}{2}\begin{bmatrix} 1 & -i & -i & -1 \\ -i & 1 & -1 & -i \\ -i & -1 & 1 & -i \\ -1 & -i & -i & 1 \end{bmatrix}$$

First we postmultiply D(0) by R:

$$D(0)R = \begin{bmatrix} 0 & 0 & 0 & 0 \\ i & 1 & -1 & i \\ i & -1 & 1 & i \\ -2 & 2i & 2i & 2 \end{bmatrix}$$

Premultiplication with R^{-1} leads to

$$D(1) = \frac{1}{4}\begin{bmatrix} 4 & -2i & -2i & 0 \\ 2i & 4 & 0 & -2i \\ 2i & 0 & 4 & -2i \\ 0 & 2i & 2i & 4 \end{bmatrix} = \frac{1}{2}\begin{bmatrix} 2 & -i & -i & 0 \\ i & 2 & 0 & -i \\ i & 0 & 2 & -i \\ 0 & i & i & 2 \end{bmatrix} \quad \text{(I.63)}$$

We note that the 90° pulse equalizes the populations and creates single-quantum coherences. The longitudinal magnetization is null while

$$\begin{aligned} M_{TA} &= -(M_o / 2)(i/1 + i/2) = -iM_o / 2 \\ M_{TX} &= -(M_o / 2)(i/1 + i/2) = -iM_o / 2 \\ M_T &= M_{TA} + M_{TX} = -iM_o \end{aligned}$$

We also note that the transverse magnetization is imaginary. Considering that $M_T = M_x + iM_y$, it follows that $M_x = 0$ and $M_y = -M_o$.

So far, the vector representation would have been much simpler to use. Let us see, though, what happens as we proceed.

4.3 Evolution from $t(1)$ to $t(2)$

The standard formula describing the (laboratory frame) time evolution of the density matrix elements in the absence of a pulse is:

$$d_{mn}(t) = d_{mn}(0)\exp(-i\omega_{mn}t) \quad \text{(I.64)}$$

d_{mn} is the matrix element and $\omega_{mn} = (E_m - E_n)/\hbar$ is the angular frequency of transition m→n. Note that $d_{mn}(0)$ is the starting point of the evolution immediately after a given pulse. In our case the elements $d_{mn}(0)$ are those of $D(1)$. If the evolution is described in a frame rotating at transmitter frequency ω_{tr} equation (I.64) becomes:

$$d_{mn}(t) = d_{mn}(0)\exp(-i\omega_{mn}t)\exp[i(m_m - m_n)\omega_{tr}t] \quad \text{(I.65)}$$

where m_m and m_n are the total magnetic quantum numbers of states m and n (see Appendix B).

Let us apply (I.65) to our particular case ($m_1 = 1$; $m_2 = m_3 = 0$; $m_4 = -1$). As expected, the diagonal elements are invariant during evolution since both exponentials are equal to 1. All single quantum coherences above diagonal have $m_m - m_n = 1$. Hence,

$$\begin{aligned} d_{mn}(t) &= d_{mn}(0)\exp(-i\omega_{mn}t)\exp(i\omega_{tr}t) \\ &= d_{mn}(0)\exp[-i(\omega_{mn} - \omega_{tr})t] \\ &= d_{mn}(0)\exp(-i\Omega_{mn}t) \end{aligned} \quad \text{(I.66)}$$

where Ω_{mn} is the evolution frequency in the rotating frame.

The rotating frame treatment is useful not only for better visualization of the vector evolution but, also, because the detection is actually made at the resulting low (audio) frequencies.

For the double-quantum coherence matrix element

$$\begin{aligned} d_{14}(t) &= d_{14}(0)\exp(-i\omega_{14}t)\exp[i(1+1)\omega_{tr}t] \\ &= d_{14}(0)\exp(-i\Omega_{14}t) \end{aligned} \quad \text{(I.67)}$$

where $\Omega_{14} = \omega_{14} - 2\omega_{tr}$. We note that both single- and double-quantum coherences evolve at low frequencies in the rotating frame.

The zero-quantum coherence matrix element is not affected by the rotating frame ($m_2 - m_3 = 0$):

$$d_{23}(t) = d_{23}(0)\exp(-i\omega_{23}t) = d_{23}(0)\exp(-i\Omega_{23}t) \quad \text{(I.68)}$$

The zero-quantum coherence evolves at low frequency in both the laboratory and rotating frame.

We now want to calculate $D(2)$, i.e., the evoluition during the first delay τ. For instance

$$d_{12}(2) = d_{12}(1)\exp(-i\Omega_{12}\tau) = -(-i/2)\exp(-i\Omega_{12}\tau) \quad \text{(I.69)}$$

To save space we let $(-i/2)\exp(-i\Omega_{mn}\tau) = B_{mn}$ and at $t(2)$ we have:

$$D(2) = \begin{bmatrix} 1 & B_{12} & B_{13} & 0 \\ B_{12}^* & 1 & 0 & B_{24} \\ B_{13}^* & 0 & 1 & B_{34} \\ 0 & B_{24}^* & B_{34}^* & 1 \end{bmatrix} \quad \text{(I.70)}$$

The z-magnetization is still zero (relaxation effects are neglected). The transverse magnetization components are:

$$M_{TA} = -(M_o/2)(B_{12}^* + B_{34}^*)$$
$$= -(iM_o/4)[\exp(i\Omega_{12}\tau) + \exp(i\Omega_{34}\tau)] \qquad (I.71)$$
$$M_{TX} = -(M_o/2)(B_{13}^* + B_{24}^*)$$
$$= -(iM_o/4)[\exp(i\Omega_{13}\tau) + \exp(i\Omega_{24}\tau)] \qquad (I.72)$$

We see that there are four vectors rotating with four different angular velocities in the equatorial (xy) plane. We can identify (see Figure I.8):

$$\Omega_{12} = \omega_{12} - \omega_{tr} = \Omega_A + \pi J$$
$$\Omega_{34} = \Omega_A - \pi J$$
$$\Omega_{13} = \Omega_X + \pi J$$
$$\Omega_{24} = \Omega_X - \pi J$$

4.4 The Second Pulse

The rotation operator for this pulse is

$$R_{180yAX} = \begin{bmatrix} 0 & 0 & 0 & 1 \\ 0 & 0 & -1 & 0 \\ 0 & -1 & 0 & 0 \\ 1 & 0 & 0 & 0 \end{bmatrix} = R_{180yAX}^{-1} \qquad (I.73)$$

At time $t(3)$ the density matrix is:

$$D(3) = R^{-1}D(2)R = \begin{bmatrix} 1 & -B_{34}^* & -B_{24}^* & 0 \\ -B_{34} & 1 & 0 & -B_{13}^* \\ -B_{24} & 0 & 1 & -B_{12}^* \\ 0 & -B_{13} & -B_{12} & 1 \end{bmatrix} \qquad (I.74)$$

Two important changes have been induced by the 180° pulse. First, all single quantum coherences were conjugated and changed sign.

This means that all x-components changed sign while the y-components remained unchanged:

$$M_T = M_x + iM_y$$

$$-M_T^* = -M_x + iM_y$$

This shows, indeed, that all four vectors rotated 180° around the y-axis. Second, coherences corresponding to fast precessing nuclei were transferred in "slots" corresponding to slow evolution. This means the vectors also changed labels.

4.5 Evolution from $t(3)$ to $t(4)$

According to (I.66) and (I.74), the evolution during the second τ delay leads to

$$\begin{aligned} d_{12}(4) &= d_{12}(3)\exp(-i\Omega_{12}\tau) = B_{34}^{*}\exp(-i\Omega_{12}\tau) \\ &= (-i/2)\exp(+i\Omega_{34}\tau)\exp(-i\Omega_{12}\tau) \\ &= (-i/2)\exp[-i(\Omega_{12}-\Omega_{34})\tau] = (-i/2)\exp(-i2\pi J\tau) \end{aligned} \quad \text{(I.75)}$$

$$d_{13}(4) = (-i/2)\exp(-i2\pi J\tau) = d_{12}(4) = U \quad \text{(I.76)}$$

$$d_{24}(4) = d_{34}(4) = (-i/2)\exp(+i2\pi J\tau) = V \quad \text{(I.77)}$$

Hence,

$$D(4) = \begin{bmatrix} 1 & U & U & 0 \\ U^* & 1 & 0 & V \\ U^* & 0 & 1 & V \\ 0 & V^* & V^* & 1 \end{bmatrix} \quad \text{(I.78)}$$

Using (I.62) we calculate the corresponding magnetization vectors:

$$\begin{aligned} M_{zA} &= M_{zX} = 0 \\ M_{TA} &= M_{TX} = -M_o(U^* + V^*) = -iM_o\cos 2\pi J\tau \end{aligned} \quad \text{(I.79)}$$

We see that while the chemical shifts refocused the coupling continues to be expressed, due to the label change.

4.6 The Third Pulse

We apply to $D(4)$ the same rotation operators we used for the first pulse and we obtain:

$$D(5)=\begin{bmatrix} 1+c & 0 & 0 & -is \\ 0 & 1 & 0 & 0 \\ 0 & 0 & 1 & 0 \\ is & 0 & 0 & 1-c \end{bmatrix} \qquad (I.80)$$

where $c = \cos 2\pi J\tau$ and $s = \sin 2\pi J\tau$.

$D(5)$ tells us that all single-quantum coherences vanished, a double-quantum coherence was created and the only existing magnetization is along the z-axis. Turning to vector representation, it is seen that before the pulse [see(I.78)] we had magnetization components on both x and y axes, since U and V are complex quantities. The vector description would indicate that the $90x$ pulse leaves the x components unchanged. In reality, as seen from the DM treatment, this does not happen since all transverse components vanish.

4.7 Evolution from $t(5)$ to $t(6)$

The double-quantum coherence element, $-is$, evolves according to (I.67):

$$D(6)=\begin{bmatrix} 1+c & 0 & 0 & w \\ 0 & 1 & 0 & 0 \\ 0 & 0 & 1 & 0 \\ w^* & 0 & 0 & 1-c \end{bmatrix} \qquad (I.81)$$

where $w = -is\exp(-i\Omega_{14}\Delta) = -i\sin 2\pi J\tau \exp(-i\Omega_{14}\Delta)$.

Our interest is in the double-quantum coherence w. In order to maximize it, we select $\tau = (2k+1)/4J$ where k = integer. Then $c = 0$ and $s = \sin[(2k+1)\pi/2] = \pm 1 = (-1)^k$

With this value of τ.

$$D(6) = \begin{bmatrix} 1 & 0 & 0 & w \\ 0 & 1 & 0 & 0 \\ 0 & 0 & 1 & 0 \\ w^* & 0 & 0 & 1 \end{bmatrix} \tag{I.82}$$

where $w = -i(-1)^k \exp(i\Omega_{14}\Delta)$

The last expression of $D(6)$ tells us that at this stage there is no magnetization at all in any of the three axes. This would be impossible to derive from the vector representation, which also could not explain the reapparition of the observable magnetization components after the fourth pulse.

4.8 The Fourth Pulse

This pulse is phase cycled, i.e., it is applied successively in various combinations along the x, y, $-x$, and $-y$ axes. The general expression of the $90\Phi AX$ operator is given in Appendix C [see(C39)].

$$R_{90\Phi AX} = \frac{1}{2}\begin{bmatrix} 1 & a & a & a^2 \\ -a^* & 1 & -1 & a \\ -a^* & -1 & 1 & a \\ a^{*2} & -a^* & -a^* & 1 \end{bmatrix} \tag{I.83}$$

where $a = i\exp(-i\Phi)$ and Φ is the angle between the x-axis and the direction of B_1. When Φ takes the value 0, 90°, 180° or 270°, the pulse is applied on axis x, y, $-x$, or $-y$, respectively.

For clarity we will discuss the *coupled* and the *noncoupled (isolated)* carbon situations separately. In the first case (coupled ^{13}C spins), we observe that at $t(6)$ [see(I.82)] the populations are equalized and all information is contained in the w elements.

The result of $R_{90\Phi AX}^{-1} D(6) R_{90\Phi AX}$ is:

$$D(7) = \begin{bmatrix} 1+a^2F & -a*G & -a*G & F \\ aG & 1-a^2F & -a^2F & a*G \\ aG & -a^2F & 1-a^2F & a*G \\ F & -aG & -aG & 1+a^2F \end{bmatrix} \tag{I.84}$$

where $$F = (w+w^*)/4 = -(1/2)(-1)^k \sin \Omega_{14}\Delta \tag{I.85}$$

$$G = (w-w^*)/4 = -(1/2)(-1)^k \cos \Omega_{14}\Delta$$

$D(7)$ shows that the newly created single quantum coherences contain the double quantum coherence information, Ω_{14}. The transverse magnetization is zero (fast and slow vectors are equal and opposite). A longitudinal magnetization proportional to $\sin\Omega_{14}$ appears. None of these could be deduced from the vector representation. Yet, the density matrix would allow the reader to draw the corresponding vectors.

To save time and space we will treat the isolated (uncoupled) carbons as an AX system in which A and X belong to two different molecules. We can use (I.81) letting $J = 0$ $(c = 1;\ s = 0;\ w = 0)$:

$$D'(6) = \begin{bmatrix} 2 & 0 & 0 & 0 \\ 0 & 1 & 0 & 0 \\ 0 & 0 & 1 & 0 \\ 0 & 0 & 0 & 0 \end{bmatrix} \tag{I.86}$$

In this case $D(7)$ becomes

$$D'(7) = \begin{bmatrix} 1 & a/2 & a/2 & 0 \\ a^*/2 & 1 & 0 & a/2 \\ a^*/2 & 0 & 1 & a/2 \\ 0 & a^*/2 & a^*/2 & 1 \end{bmatrix} \tag{I.87}$$

$D'(7)$ shows only single quantum coherences and equalized populations. The transverse magnetization of the noncoupled spins is equal to their equilibrium magnetization M'_o (its orientation depends on the value of Φ).

4.9 Detection

The magnetization at $t(8)$, due to coupled (cpl) and noncoupled (ncpl) nuclei can be calculated starting from the single quantum coherences in (I.84) and (I.87) respectively.

$$M_T(cpl) = M[\exp(i\Omega_{12}t_d) + \exp(i\Omega_{13}t_d) - \exp(i\Omega_{24}t_d) - \exp(i\Omega_{34}t_d)] \qquad (I.88)$$

where $\quad M = -M_o aG^* = (M_0 / 2)\exp(-i\Phi)(-1)^k \cos\Omega_{14}\Delta$

These are the four peaks of the coupled AX system shown as a schematic contour plot in Figure I.9.

For the noncoupled carbons

$$M_T(ncpl) = M'[2\exp(i\Omega_A t_d) + 2\exp(i\Omega_X t_d)] \qquad (I.89)$$

where $\quad M' = -M_o' a^* /2 = (-iM_o' / 2)\exp(-i\Phi)$

These are the two peaks of the uncoupled nuclei. Each of them is 200 times more intense than each of the four peaks in (I.88).

The culminating point of INADEQUATE is the selective detection of $M_T(cpl)$. We note that $M_T(cpl)$ and $M_T(ncpl)$ depend in opposite ways on Φ since one contains a and the other a^* [cf.(I.88) and (I.89)]. They can be discriminated by cycling Φ and properly choosing the receiver phase Ψ [the detected signal $S = M_T \exp(-i\Psi)$]. The table below shows that two cycles are sufficient to eliminate $M_T(ncpl)$.

Cycle	Φ	Ψ	ncpl -------------- exp$i(\Phi-\Psi)$	cpl -------------- exp$i(-\Phi-\Psi)$
1	0	0	1	1
2	90°	-90°	-1	1
(1+2)			0	2

Even small imperfections of the rf pulse will allow leakage of the strong undesired signal into the resultant spectrum. This makes it necessary to apply one of several cycling patterns consisting of up to

256 steps, which attempt to cancel the effects of too long, too short, or incorrectly phased, pulses.

4.10 Carbon-Carbon Connectivity

A significant extension of INADEQUATE is its adaptation for two-dimensional experiments. The second time-domain (in addition to t_d) is created by making Δ variable. All we have to do now is to discuss (I.88) and (I.89) in terms of two time variables.

As seen in Figure I.9, all four peaks of $M_T(cpl)$ will be aligned in domain Δ along the same frequency $\Omega_{14} = \Omega_A + \Omega_X$. The great advantage of the 2D display consists of the fact that every pair of coupled carbons will exhibit its pair of doublets along its own Ω_{14} frequency. This allows us to trace out the carbon skeleton of an organic molecule.

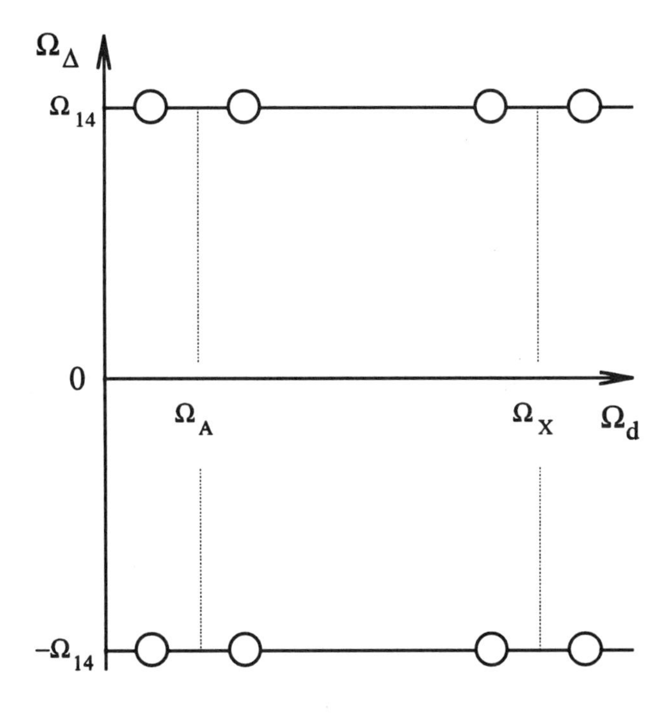

Figure I.9. The four peaks due to a pair of coupled ^{13}C atoms. The vertical scale is twice larger than the horizontal scale.

The student is invited to identify the molecule whose 2D spectrum is shown in Figure I.10 and to determine its carbon-carbon connectivities. The answer is in the footnote on page 59.

It should be noted that $M_T(ncpl)$ does not depend on Δ [cf.(I.81)]. This means that, when incompletely eliminated, the peaks of isolated carbons will be "axial" (dotted circles on the zero frequency line of domain Δ).

We also note that $M_T(cpl)$ is phase modulated with respect to t_d and amplitude modulated with respect to Δ. Consequently, mirror-image peaks will appear at frequencies $-\Omega_{14}$. This reduces the intensity of the displayed signals and imposes restrictions on the choice of the transmitter frequency, increasing the size of the data matrix. A modified sequence has been proposed to obtain phase modulation with respect to Δ, the analog of a quadrature detection in domain Δ.

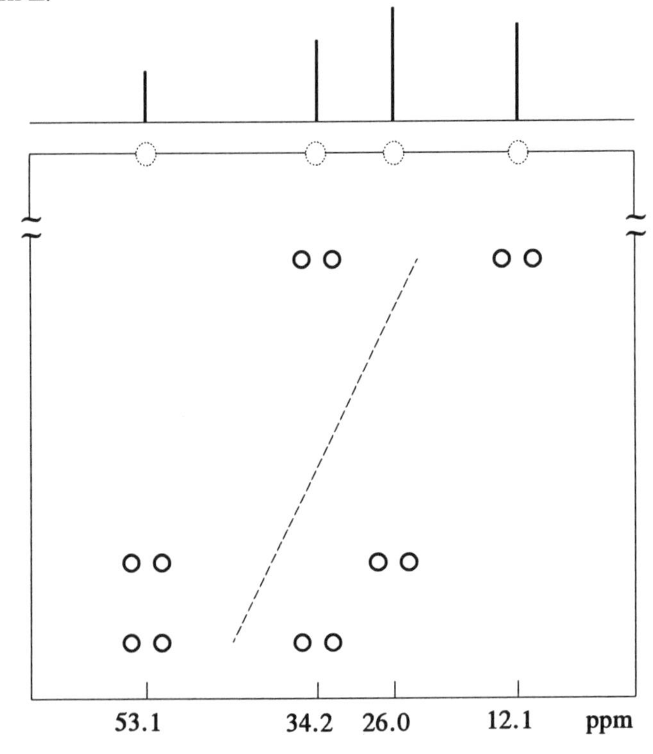

Figure I.10. The carbon-carbon connectivity spectrum of a molecule with MW = 137.

5. DENSITY MATRIX DESCRIPTION OF COSY (HOMONUCLEAR CORRELATION SPECTROSCOPY)

COSY (COrrelation SpectroscopY) is widely used, particularly for disentangling complicated proton spectra by proton-proton chemical shift correlation and elucidation of the coupling pattern. Of course, other spin 1/2 systems such as ^{19}F can be successfully studied with this 2D sequence. The basic sequence is shown in Figure I.11.

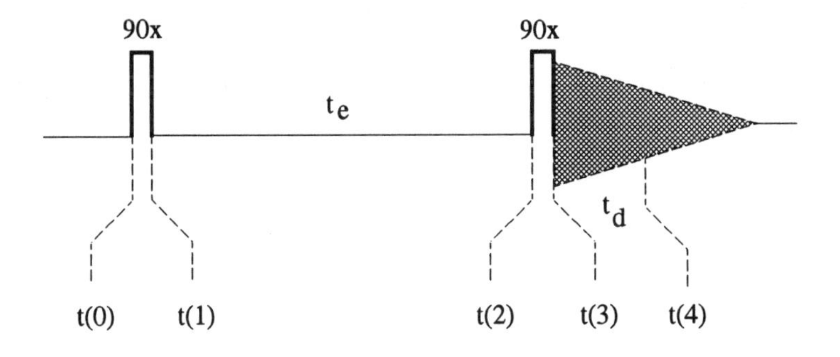

Figure I.11. The Basic COSY sequence (without phase cycling): $90x - t_e - 90x - \mathrm{AT}$

5.1 Equilibrium Populations

Since we deal with a homonuclear AX system, the populations at thermal equilibrium follow a pattern identical to that of INADEQUATE (see I.60 and I.61). The initial density matrix is:

$$D(0) = \begin{bmatrix} 0 & 0 & 0 & 0 \\ 0 & 1 & 0 & 0 \\ 0 & 0 & 1 & 0 \\ 0 & 0 & 0 & 2 \end{bmatrix} \tag{I.90}$$

5.2 The First Pulse

Here, also, we can use the results from INADEQUATE (I.63) since the first pulse is a nonselective 90*xAX*:

$$D(1)=\frac{1}{2}\begin{bmatrix} 2 & -i & -i & 0 \\ i & 2 & 0 & -i \\ i & 0 & 2 & -i \\ 0 & i & i & 2 \end{bmatrix} \tag{I.91}$$

5.3 Evolution from $t(1)$ to $t(2)$

Only nondiagonal terms are affected by evolution. The single-quantum coherences will evolve, each with its own angular frequency, leading to:

$$D(2)=\begin{bmatrix} 1 & A & B & 0 \\ A^* & 1 & 0 & C \\ B^* & 0 & 1 & D \\ 0 & C^* & D^* & 1 \end{bmatrix} \tag{I.92}$$

where

$$\begin{aligned} A&=(-i/2)\exp(-i\Omega_{12}t_e) \\ B&=(-i/2)\exp(-i\Omega_{13}t_e) \\ C&=(-i/2)\exp(-i\Omega_{24}t_e) \\ D&=(-i/2)\exp(-i\Omega_{34}t_e) \end{aligned} \tag{I.93}$$

5.4 The Second Pulse

To calculate $D(3)=R^{-1}D(2)R$, we use the rotation operators for the 90*xAX* pulse given in (I.34)

$$R=\frac{1}{2}\begin{bmatrix} 1 & i & i & -1 \\ i & 1 & -1 & i \\ i & -1 & 1 & i \\ -1 & i & i & 1 \end{bmatrix} \quad ; \quad R^{-1}=\frac{1}{2}\begin{bmatrix} 1 & -i & -i & -1 \\ -i & 1 & -1 & -i \\ -i & -1 & 1 & -i \\ -1 & -i & -i & 1 \end{bmatrix}$$

The postmultiplication $D(2)R$ yields

$$\frac{1}{2}\begin{bmatrix} 1+iA+iB & i+A-B & i-A+B & -1+iA+iB \\ i+A^*-C & 1+iA^*+iC & -1+iA^*+iC & i-A^*+C \\ i+B^*-D & -1+iB^*+iD & 1+iB^*+iD & i-B^*+D \\ -1+iC^*+iD^* & i+C^*-D^* & i-C^*+D^* & 1+iC^*+iD^* \end{bmatrix}$$

We then premultiply this result by R^{-1} and obtain

$$D(3) = R^{-1}D(2)R$$

$$= \frac{1}{4}\begin{bmatrix} \begin{matrix}4+iA-iA^*\\+iB-iB^*\\+iC-iC^*\\+iD-iD^*\end{matrix} & \begin{matrix}+A+A^*\\-B+B^*\\+C-C^*\\+D+D^*\end{matrix} & \begin{matrix}-A+A^*\\+B+B^*\\+C+C^*\\+D-D^*\end{matrix} & \begin{matrix}+iA+iA^*\\+iB+iB^*\\-iC-iC^*\\-iD-iD^*\end{matrix} \\ \\ \begin{matrix}+A+A^*\\+B-B^*\\-C+C^*\\+D+D^*\end{matrix} & \begin{matrix}4-iA+iA^*\\+iB-iB^*\\+iC-iC^*\\-iD+iD^*\end{matrix} & \begin{matrix}+iA+iA^*\\-iB-iB^*\\+iC+iC^*\\-iD-iD^*\end{matrix} & \begin{matrix}+A-A^*\\+B+B^*\\+C+C^*\\-D+D^*\end{matrix} \\ \\ \begin{matrix}+A-A^*\\+B+B^*\\+C+C^*\\-D+D^*\end{matrix} & \begin{matrix}-iA-iA^*\\+iB+iB^*\\-iC-iC^*\\+iD+iD^*\end{matrix} & \begin{matrix}4+iA-iA^*\\-iB+iB^*\\-iC+iC^*\\+iD-iD^*\end{matrix} & \begin{matrix}+A+A^*\\+B-B^*\\-C+C^*\\+D+D^*\end{matrix} \\ \\ \begin{matrix}-iA-iA^*\\-iB-iB^*\\+iC+iC^*\\+iD+iD^*\end{matrix} & \begin{matrix}-A+A^*\\+B+B^*\\+C+C^*\\+D-D^*\end{matrix} & \begin{matrix}+A+A^*\\-B+B^*\\+C-C^*\\+D+D^*\end{matrix} & \begin{matrix}4-iA+iA^*\\-iB+iB^*\\-iC+iC^*\\-iD+iD^*\end{matrix} \end{bmatrix}$$

(I.94)

One can check that the population sum $d_{11} + d_{22} + d_{33} + d_{44}$ (trace of the matrix) is invariant, i.e., it has the same value for $D(0)$ through $D(3)$. Also, $D(3)$ is Hermitian (the density matrix always is). In doing this verification we keep in mind that the sums $A + A^*$, $B + B^*$, etc., are all *real* quantities, while the differences $A - A^*$, $B - B^*$, etc., are *imaginary*. This can be used to simplify the expression of $D(3)$ by employing the following notations:

$$\begin{aligned} A &= (-i/2)\exp(-i\Omega_{12}t_e) = (-i/2)(\cos\Omega_{12}t_e - i\sin\Omega_{12}t_e) = \\ &= (-i/2)(c_{12} - is_{12}) = -(1/2)(s_{12} + ic_{12}) \\ A^* &= -(1/2)(s_{12} - ic_{12}) \end{aligned}$$

With similar notations for B, C, and D we obtain:

$$\begin{aligned} A &= -(1/2)(s_{12} + ic_{12}) \;;\; A + A^* = -s_{12} \;;\; A - A^* = -ic_{12} \\ B &= -(1/2)(s_{13} + ic_{13}) \;;\; B + B^* = -s_{13} \;;\; B - B^* = -ic_{13} \\ C &= -(1/2)(s_{24} + ic_{24}) \;;\; C + C^* = -s_{24} \;;\; C - C^* = -ic_{24} \\ D &= -(1/2)(s_{34} + ic_{34}) \;;\; D + D^* = -s_{34} \;;\; D - D^* = -ic_{34} \end{aligned} \tag{I.95}$$

With the new sine/cosine notations, $D(3)$ becomes:

$$\frac{1}{4}\begin{bmatrix} 4 + c_{12} + c_{13} & -s_{12} + ic_{13} & ic_{12} - s_{13} & -is_{12} - is_{13} \\ + c_{24} + c_{34} & -ic_{24} - s_{34} & -s_{24} - ic_{34} & +is_{24} + is_{34} \\ & & & \\ -s_{12} - ic_{13} & 4 - c_{12} + c_{13} & -is_{12} + is_{13} & -ic_{12} - s_{13} \\ +ic_{24} - s_{34} & + c_{24} - c_{34} & -is_{24} + is_{34} & -s_{24} + ic_{34} \\ & & & \\ -ic_{12} - s_{13} & is_{12} - is_{13} & 4 + c_{12} - c_{13} & -s_{12} - ic_{13} \\ -s_{24} + ic_{34} & +is_{24} - is_{34} & - c_{24} + c_{34} & +ic_{24} - s_{34} \\ & & & \\ is_{12} + is_{13} & ic_{12} - s_{13} & -s_{12} + ic_{13} & 4 - c_{12} - c_{13} \\ -is_{24} - is_{34} & -s_{24} - ic_{34} & -ic_{24} - s_{34} & - c_{24} - c_{34} \end{bmatrix} \tag{I.96}$$

5.5 Detection

Since no other pulse follows after $t(3)$ we will consider only the evolution of the observable (single quantum) elements d_{12}, d_{34}, d_{13}, and d_{24} in the time domain t_d. Before evolution they are (from I.96):

$$\begin{aligned} d_{12}(3) &= (i/4)(+is_{12} + is_{34} + c_{13} - c_{24}) \\ d_{34}(3) &= (i/4)(+is_{12} + is_{34} - c_{13} + c_{24}) \\ d_{13}(3) &= (i/4)(+c_{12} - c_{34} + is_{13} + is_{24}) \\ d_{24}(3) &= (i/4)(-c_{12} + c_{34} + is_{13} + is_{24}) \end{aligned} \tag{I.97}$$

Their complex conjugates, which are needed for the calculation of magnetization components, are:

$$\begin{aligned} d^*_{12}(3) &= (i/4)(+is_{12} + is_{34} - c_{13} + c_{24}) \\ d^*_{34}(3) &= (i/4)(+is_{12} + is_{34} + c_{13} - c_{24}) \\ d^*_{13}(3) &= (i/4)(-c_{12} + c_{34} + is_{13} + is_{24}) \\ d^*_{24}(3) &= (i/4)(+c_{12} - c_{34} + is_{13} + is_{24}) \end{aligned} \tag{I.98}$$

Each of the four matrix elements above contains all four frequencies evolving in domain t_e, namely: $\Omega_{12}t_e$, $\Omega_{34}t_e$, $\Omega_{13}t_e$, and $\Omega_{24}t_e$. During detection, each of them will evolve with its own frequency in the domain t_d:

$$\begin{aligned} d^*_{12}(4) &= (i/4)(+is_{12} + is_{34} - c_{13} + c_{24})\exp(i\Omega_{12}t_d) \\ d^*_{34}(4) &= (i/4)(+is_{12} + is_{34} + c_{13} - c_{24})\exp(i\Omega_{34}t_d) \\ d^*_{13}(4) &= (i/4)(-c_{12} + c_{34} + is_{13} + is_{24})\exp(i\Omega_{13}t_d) \\ d^*_{24}(4) &= (i/4)(+c_{12} - c_{34} + is_{13} + is_{24})\exp(i\Omega_{24}t_d) \end{aligned} \tag{I.99}$$

Expression (I.99) shows that each t_d frequency is modulated by all four t_e frequencies. Thus, we expect sixteen peaks in the 2D plot. Actually, there will be 32 peaks, because only the t_d domain is phase modulated, while the t_e domain is amplitude modulated (i.e., it contains sine/cosine expressions).

Each sine or cosine implies both the positive and the negative frequency according to

$$c_{jk} = \cos \Omega_{jk} t_e = \frac{1}{2}\left[\exp\left(i\Omega_{jk} t_e\right) + \exp\left(-i\Omega_{jk} t_e\right)\right]$$

$$is_{jk} = i \sin \Omega_{jk} t_e = \frac{1}{2}\left[\exp\left(i\Omega_{jk} t_e\right) - \exp\left(-i\Omega_{jk} t_e\right)\right]$$

This leads to the 32 peak contour plot in Figure I.12.

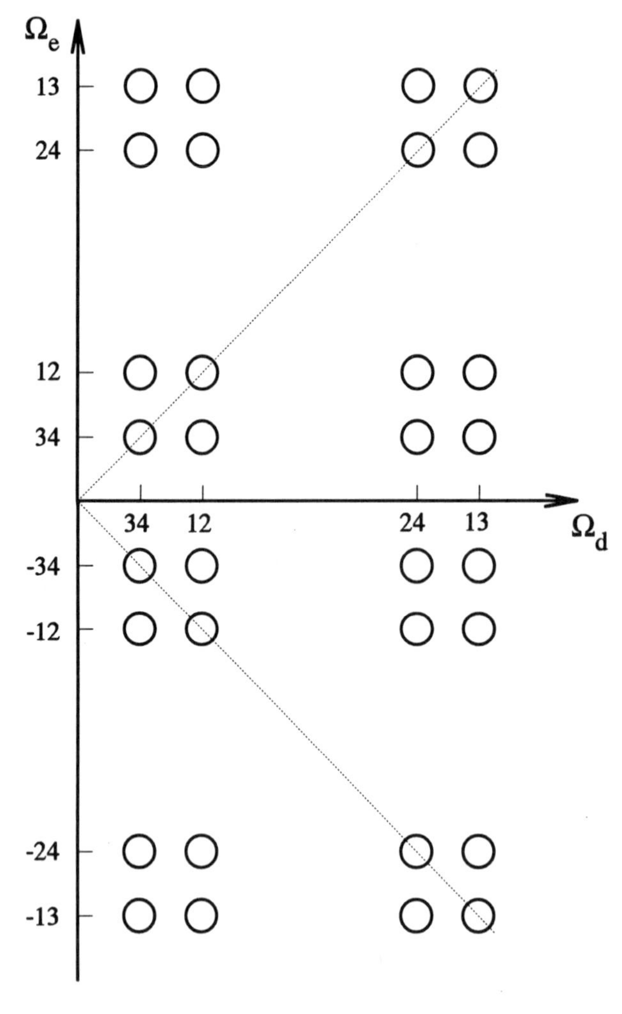

Figure I.12. Contour plot of COSY without phase cycling. The transmitter frequency is on one side of the spectrum.

We can plot the positive frequencies only, but the amplitude modulation in domain t_e still is a major drawback since it requires placing the transmitter frequency outside the spectrum (i.e., we lose the advantage of quadrature detection in both domains). The spectral widths have to be doubled and the data matrix increases by a factor of four.

If the transmitter is placed within the spectrum (e.g., between Ω_{12} and Ω_{24}), a messy pattern is obtained as shown in Figure I.13. The next section shows how this can be circumvented by means of an appropriate phase cycling.

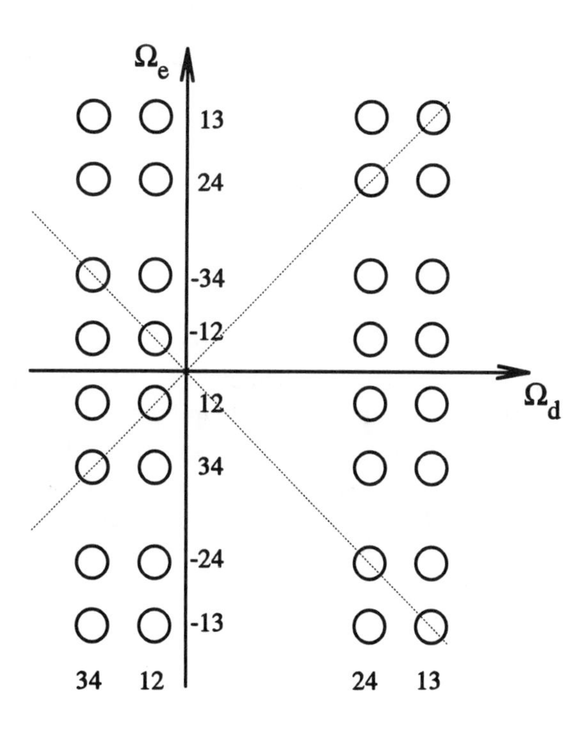

Figure I.13. Contour plot of COSY without phase cycling. If the transmitter is placed within the spectrum, it causes overlap of positive and negative frequencies.

6. COSY WITH PHASE CYCLING

6.1 Comparison with the Previous Sequence

The sequence for COSY with phase cycling shown in Figure I.14 differs from that discussed above (Figure I.11) only by the cycling of the second pulse.

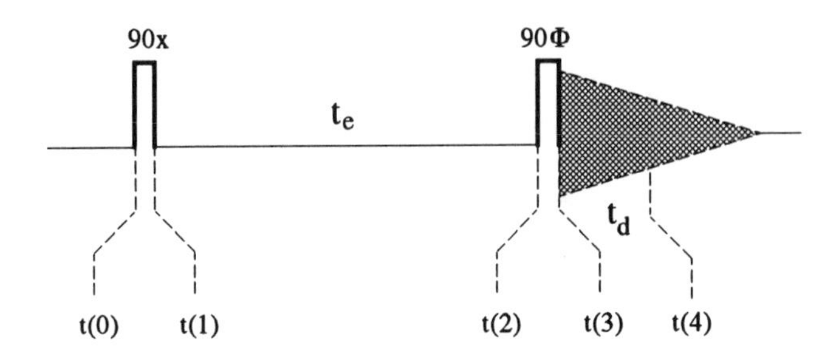

Figure I.14. COSY sequence with phase cycling of second pulse: $90x - t_e - 90\Phi - \mathrm{AT}$

Moreover, only two steps are theoretically necessary to eliminate negative frequencies in domain t_e. The second pulse is successively phased in x and y. Rather than doing the density matrix calculations for an arbitrary phase Φ, we will take advantage of the fact that we have already treated the x phase in the previous section. Only the effect of $90yAX$ must then be calculated (see Figure I.15).

Since the two sequences in Figures I.11 and I.15 have a common segment [$t(0)$ to $t(2)$], we can take $D(2)$ from the previous section [see (I.92). and (I.93)].

$$D(2) = \begin{bmatrix} 1 & A & B & 0 \\ A^* & 1 & 0 & C \\ B^* & 0 & 1 & D \\ 0 & C^* & D^* & 1 \end{bmatrix} \qquad \text{(I.100)}$$

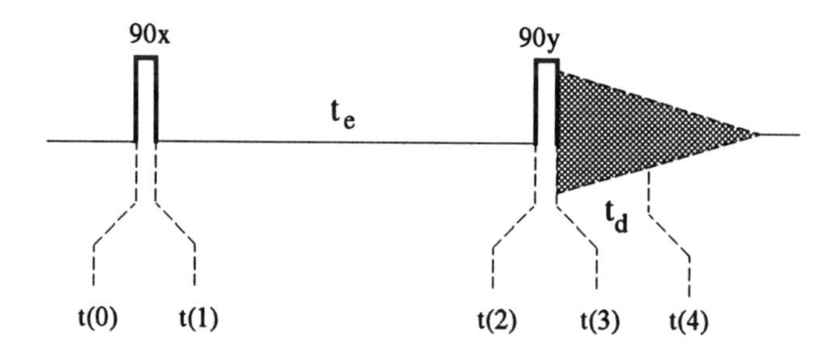

Figure I.15. The second step of the phase cycled COSY sequence: $90x - t_e - 90y - AT$

6.2 The Second Pulse

The rotation operator for the $90yAX$ pulse can be obtained by multiplying R_{90yA} by R_{90yX}. These operators are (see Appendix C):

$$R_{90yA} = \frac{1}{\sqrt{2}}\begin{bmatrix} 1 & 1 & 0 & 0 \\ -1 & 1 & 0 & 0 \\ 0 & 0 & 1 & 1 \\ 0 & 0 & -1 & 1 \end{bmatrix} \quad ; \quad R_{90yX} = \frac{1}{\sqrt{2}}\begin{bmatrix} 1 & 0 & 1 & 0 \\ 0 & 1 & 0 & 1 \\ -1 & 0 & 1 & 0 \\ 0 & -1 & 0 & 1 \end{bmatrix}$$

The result of the multiplication, $R = R_{90yAX}$, is shown below together with its reciprocal, R^{-1}.

$$R = \frac{1}{2}\begin{bmatrix} 1 & 1 & 1 & 1 \\ -1 & 1 & -1 & 1 \\ -1 & -1 & 1 & 1 \\ 1 & -1 & -1 & 1 \end{bmatrix} \quad ; \quad R^{-1} = \frac{1}{2}\begin{bmatrix} 1 & -1 & -1 & 1 \\ 1 & 1 & -1 & -1 \\ 1 & -1 & 1 & -1 \\ 1 & 1 & 1 & 1 \end{bmatrix} \qquad \text{(I.101)}$$

The postmultiplication $D(2)R$ yields

$$\frac{1}{2}\begin{bmatrix} 1-A-B & 1+A-B & 1-A+B & 1+A+B \\ -1+A*+C & 1+A*-C & -1+A*-C & 1+A*+C \\ -1+B*+D & -1+B*-D & 1+B*-D & i+B*+D \\ 1-C*-D* & -1+C*-D* & -1-C*+D* & 1+C*+D* \end{bmatrix}$$

Premultiplying this result by R^{-1} gives

$$D(3) = R^{-1}D(2)R =$$

$$\frac{1}{4}\begin{bmatrix} \begin{array}{l} 4-A-A* \\ -B-B* \\ -C-C* \\ -D-D* \end{array} & \begin{array}{l} +A-A* \\ -B-B* \\ +C+C* \\ +D-D* \end{array} & \begin{array}{l} -A-A* \\ +B-B* \\ +C-C* \\ +D+D* \end{array} & \begin{array}{l} +A-A* \\ +B-B* \\ -C+C* \\ -D+D* \end{array} \\ \\ \begin{array}{l} -A+A* \\ -B-B* \\ +C+C* \\ -D+D* \end{array} & \begin{array}{l} 4+A+A* \\ -B-B* \\ -C-C* \\ +D+D* \end{array} & \begin{array}{l} -A+A* \\ +B-B* \\ -C+C* \\ +D-D* \end{array} & \begin{array}{l} A+A* \\ +B-B* \\ +C-C* \\ -D-D* \end{array} \\ \\ \begin{array}{l} -A-A* \\ -B+B* \\ -C+C* \\ +D+D* \end{array} & \begin{array}{l} +A-A* \\ -B+B* \\ +C-C* \\ -D+D* \end{array} & \begin{array}{l} 4-A-A* \\ +B+B* \\ +C+C* \\ -D-D* \end{array} & \begin{array}{l} +A-A* \\ +B+B* \\ -C-C* \\ +D-D* \end{array} \\ \\ \begin{array}{l} -A+A* \\ -B+B* \\ +C-C* \\ +D-D* \end{array} & \begin{array}{l} +A+A* \\ -B+B* \\ -C+C* \\ -D-D* \end{array} & \begin{array}{l} -A+A* \\ +B+B* \\ -C-C* \\ -D+D* \end{array} & \begin{array}{l} 4+A+A* \\ +B+B* \\ +C+C* \\ +D+D* \end{array} \end{bmatrix} \tag{I.102}$$

With the same notations as in (I.95) :

$$
\begin{aligned}
A &= -(1/2)(s_{12} + ic_{12}) \ ; \ A + A^* = -s_{12} \ ; \ A - A^* = -ic_{12} \\
B &= -(1/2)(s_{13} + ic_{13}) \ ; \ B + B^* = -s_{13} \ ; \ B - B^* = -ic_{13} \\
C &= -(1/2)(s_{24} + ic_{24}) \ ; \ C + C^* = -s_{24} \ ; \ C - C^* = -ic_{24} \\
D &= -(1/2)(s_{34} + ic_{34}) \ ; \ D + D^* = -s_{34} \ ; \ D - D^* = -ic_{34}
\end{aligned} \tag{I.103}
$$

$D(3)$ becomes

$$
\frac{1}{4}\begin{bmatrix}
\begin{matrix}4 + s_{12} + s_{13} \\ + s_{24} + s_{34}\end{matrix} & \begin{matrix}-ic_{12} + s_{13} \\ -s_{24} - ic_{34}\end{matrix} & \begin{matrix}+s_{12} - ic_{13} \\ -ic_{24} - s_{34}\end{matrix} & \begin{matrix}-ic_{12} - ic_{13} \\ +ic_{24} + ic_{34}\end{matrix} \\
\begin{matrix}+ic_{12} + s_{13} \\ -s_{24} + ic_{34}\end{matrix} & \begin{matrix}4 - s_{12} + s_{13} \\ + s_{24} - s_{34}\end{matrix} & \begin{matrix}+ic_{12} - ic_{13} \\ +ic_{24} - ic_{34}\end{matrix} & \begin{matrix}-s_{12} - ic_{13} \\ -ic_{24} + s_{34}\end{matrix} \\
\begin{matrix}+s_{12} + ic_{13} \\ +ic_{24} - s_{34}\end{matrix} & \begin{matrix}-ic_{12} + ic_{13} \\ -ic_{24} + ic_{34}\end{matrix} & \begin{matrix}4 + s_{12} - s_{13} \\ - s_{24} + s_{34}\end{matrix} & \begin{matrix}-ic_{12} - s_{13} \\ +s_{24} - ic_{34}\end{matrix} \\
\begin{matrix}+ic_{12} + ic_{13} \\ -ic_{24} - ic_{34}\end{matrix} & \begin{matrix}-s_{12} + ic_{13} \\ +ic_{24} + s_{34}\end{matrix} & \begin{matrix}+ic_{12} - s_{13} \\ +s_{24} + ic_{34}\end{matrix} & \begin{matrix}4 - s_{12} - s_{13} \\ - s_{24} - s_{34}\end{matrix}
\end{bmatrix} \tag{I.104}
$$

We have to consider only the observable matrix elements :

$$
\begin{aligned}
d_{12}(3) &= (i/4)(-c_{12} - c_{34} - is_{13} + is_{24}) \\
d_{34}(3) &= (i/4)(-c_{12} - c_{34} + is_{13} - is_{24}) \\
d_{13}(3) &= (i/4)(-is_{12} + is_{34} - c_{13} - c_{24}) \\
d_{24}(3) &= (i/4)(+is_{12} - is_{34} - c_{13} - c_{24})
\end{aligned} \tag{I.105}
$$

Comparing the results for phase y (I.105) with those for phase x (I.97) shows that c and $-is$ are interchanged in all terms. This will lead to the desired phase modulation. Addition of (I.97) and (I.105) gives

$$d_{12}(3)=(i/4)\left[(is_{12}-c_{12})+(is_{34}-c_{34})+(c_{13}-is_{13})+(-c_{24}+is_{24})\right]$$
$$d_{34}(3)=(i/4)\left[(is_{12}-c_{12})+(is_{34}-c_{34})+(-c_{13}+is_{13})+(c_{24}-is_{24})\right]$$
$$d_{13}(3)=(i/4)\left[(c_{12}-is_{12})+(-c_{34}+is_{34})+(is_{13}-c_{13})+(is_{24}-c_{24})\right]$$
$$d_{24}(3)=(i/4)\left[(-c_{12}+is_{12})+(c_{34}-is_{34})+(is_{13}-c_{13})+(is_{24}-c_{24})\right]$$

(I.106)

The complex conjugates of the matrix elements above (needed for the expression of the magnetization components) are

$$d^*_{12}(3)=(i/4)\left[(is_{12}+c_{12})+(is_{34}+c_{34})+(-c_{13}-is_{13})+(c_{24}+is_{24})\right]$$
$$d^*_{34}(3)=(i/4)\left[(is_{12}+c_{12})+(is_{34}+c_{34})+(c_{13}+is_{13})+(-c_{24}-is_{24})\right]$$
$$d^*_{13}(3)=(i/4)\left[(-c_{12}-is_{12})+(c_{34}+is_{34})+(is_{13}+c_{13})+(is_{24}+c_{24})\right]$$
$$d^*_{24}(3)=(i/4)\left[(c_{12}+is_{12})+(-c_{34}-is_{34})+(is_{13}+c_{13})+(is_{24}+c_{24})\right]$$

(I.107)

We observe that every parenthesis represents an exponential, therefore we can use the notation

$$e_{jk}=c_{jk}+is_{jk}=\exp\left(i\Omega_{jk}t_e\right) \qquad \text{(I.108)}$$

With this notation (I.107) becomes

$$d^*_{12}(3)=(i/4)\left(+e_{12}+e_{34}-e_{13}+e_{24}\right)$$
$$d^*_{34}(3)=(i/4)\left(+e_{12}+e_{34}+e_{13}-e_{24}\right)$$
$$d^*_{13}(3)=(i/4)\left(-e_{12}+e_{34}+e_{13}+e_{24}\right) \qquad \text{(I.109)}$$
$$d^*_{24}(3)=(i/4)\left(+e_{12}-e_{34}+e_{13}+e_{24}\right)$$

The reader is reminded that the above expressions represent the summation from two acquisitions, one with phase x and the other with phase y for the second pulse. Instead of sines or cosines, they contain only exponentials. In other words we have achieved phase modulation in the domain t_e.

6.3 Detection

The evolution during t_d is identical for the two acquisitions, therefore we can apply it after the summation (in I.109).

$$\begin{aligned}
d_{12}^*(4) &= (i/4)(+e_{12}+e_{34}-e_{13}+e_{24})\exp(i\Omega_{12}t_d) \\
d_{34}^*(4) &= (i/4)(+e_{12}+e_{34}+e_{13}-e_{24})\exp(i\Omega_{34}t_d) \\
d_{13}^*(4) &= (i/4)(-e_{12}+e_{34}+e_{13}+e_{24})\exp(i\Omega_{13}t_d) \\
d_{24}^*(4) &= (i/4)(+e_{12}-e_{34}+e_{13}+e_{24})\exp(i\Omega_{24}t_d)
\end{aligned} \tag{I.110}$$

The expressions (I.110) correspond to a contour plot with 16 peaks, as represented in Figure I.16. This pattern is no longer dependent on the position of the transmitter since we have achieved the equivalent of quadrature detection in both domains. There are four diagonal terms, each having the same frequency in both domains t_e and t_d. If the two frequencies differ only by J (e.g., we have Ω_{12} in domain t_e and Ω_{34} in domain t_d) we have a "near-diagonal" peak. There are four such peaks. The remaining eight are referred to as *off-diagonal* or *cross-peaks* and their presence indicates that spins A and X are coupled. Two noncoupled spins will exhibit only diagonal peaks. We can verify that this is so by discussing what happens if J becomes vanishingly small. In this case $\Omega_{12} = \Omega_{34} = \Omega_A$ (center of the doublet); likewise $\Omega_{13} = \Omega_{24} = \Omega_X$. Therefore each group of 4 peaks in Figure I.16 collapses into a single peak in the center of the corresponding square. The diagonal and near-diagonal peaks of nucleus A are all positive and they collapse into one diagonal peak, four times larger. The same is true for nucleus X. The off-diagonal peaks come in groups of four, two positive and two negative. When they collapse (J=0) they cancel each other and there will be no off-diagonal peak. The net result is a spectrum with just two peaks, both on the diagonal, with frequencies Ω_A and Ω_X. The above discussion shows that COSY is not suited for spectra with ill-resolved multiplets because there will be destructive overlap in the off-diagonal groups.

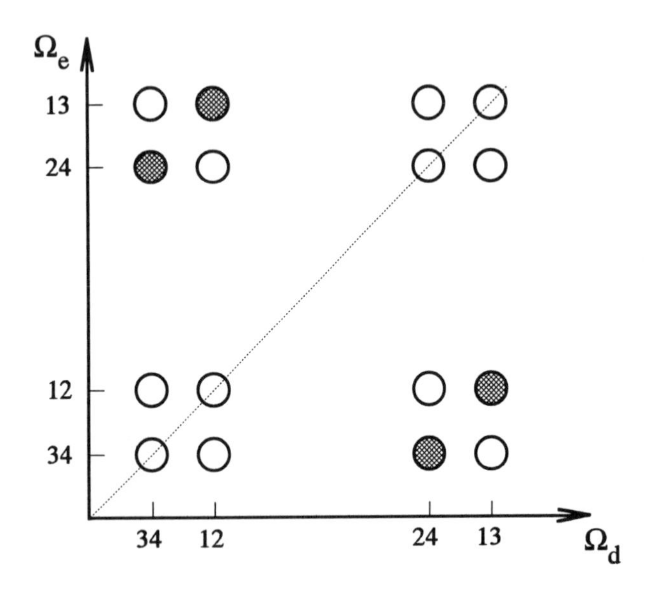

Figure I.16. Contour plot of COSY with phase cycling. Open circles are positive peaks; shaded circles are negative peaks.

There is one more observation. It is common practice to plot the 2D spectra in the "magnitude calculation" (MC) or "absolute value" mode, to avoid phasing problems. The magnitude calculation (the absolute value of a complex quantity) is performed by the computer after Fourier transform in both domains. Two peaks, one positive and one negative, will both become positive in MC, provided they are well resolved. If they are ill-resolved, they will cancel each other partially and will yield a much smaller signal. The MC is applied to this signal and it cannot represent the original amplitude of the two peaks. Therefore the use of magnitude calculation does not provide a solution for poorly resolved multiplets.

One word about the utility of phase cycling. While it is well known that this procedure helps canceling out radiofrequency interferences and pulse imperfections, we have just seen that it can be useful for other purposes. In INADEQUATE it helps eliminate the NMR signal from the ^{13}C-^{12}C pairs while preserving that from ^{13}C-^{13}C pairs. In COSY it helps achieve phase modulation in domain t_e. In both cases a two step cycling is theoretically sufficient. However, to efficiently compensate for hardware imperfections, cycling in more than two steps is generally employed.

7. CONCLUSION OF PART I

The density matrix approach described above constitutes a very clear and useful means for understanding the multipulse sequences of modern NMR. The limitation of this approach is the rapidly increasing volume of calculation with increasing number of nuclei. The size of the matrix goes from 16 elements for a two spin 1/2 system to 64 and 256 elements for three and four spin systems, respectively.

We must therefore resort to an avenue which affords a "shorthand" for the description of rotations and evolutions. The new avenue we present in the second part of this monograph is the *product operator* formalism.

Answer to the problem on page 44: The contour plot in Figure I.10 is the carbon-carbon connectivity spectrum of 2-bromobutane.

2

The Product Operator Formalism

1. INTRODUCTION

In this section we will see that the density matrix at equilibrium can be expressed in terms of the spin angular momentum component I_z of each nucleus. Moreover, effects of pulses (rotations) and evolutions of noncoupled spins can also be described in terms of angular momentum components (I_x,I_y,I_z). However, in order to express evolutions of coupled spins, we will need additional "building blocks" besides angular momentum components. We will thus introduce a "basis set" which is composed of "product operators." The latter are either products between angular momentum components or products of angular momentum components with the unit matrix. We will describe this approach and will apply it to several pulse sequences, beginning with 2DHETCOR.

2. EXPRESSING THE DENSITY MATRIX IN TERMS OF ANGULAR MOMENTUM COMPONENTS

We start with the same procedure of describing the density matrix at equilibrium, $D(0)$, as in Part I. In order to generalize the approach to an AX (not only a CH) system we preserve separate Boltzman factors, $1 + p$ and $1 + q$, for nuclei A and X, respectively [see (I.3-5)]:

$$D(0)=\frac{1}{N}\begin{bmatrix} 1 & 0 & 0 & 0 \\ 0 & 1+p & 0 & 0 \\ 0 & 0 & 1+q & 0 \\ 0 & 0 & 0 & 1+p+q \end{bmatrix} \quad \text{(II.1)}$$

N is the number of states (4 for an AX system).

As we did in Part I, we separate the unit matrix from the matrix representing population differences. However, here we choose as a factor for the unit matrix the average population $(1+p/2+q/2)/N$:

$$D(0)=\frac{1+p/2+q/2}{N}\begin{bmatrix} 1 & 0 & 0 & 0 \\ 0 & 1 & 0 & 0 \\ 0 & 0 & 1 & 0 \\ 0 & 0 & 0 & 1 \end{bmatrix}$$

$$+\frac{1}{N}\begin{bmatrix} -p/2-q/2 & 0 & 0 & 0 \\ 0 & p/2-q/2 & 0 & 0 \\ 0 & 0 & -p/2+q/2 & 0 \\ 0 & 0 & 0 & p/2+q/2 \end{bmatrix}$$

(II.2)

Again, we ignore the term containing the unit matrix which does not contribute to magnetization.

$$D(0)=\frac{1}{N}\begin{bmatrix} -p/2-q/2 & 0 & 0 & 0 \\ 0 & p/2-q/2 & 0 & 0 \\ 0 & 0 & -p/2+q/2 & 0 \\ 0 & 0 & 0 & p/2+q/2 \end{bmatrix}$$

(II.3)

Separation of the p and q terms gives

$$D(0) = -\frac{p}{2N}\begin{bmatrix} 1 & 0 & 0 & 0 \\ 0 & -1 & 0 & 0 \\ 0 & 0 & 1 & 0 \\ 0 & 0 & 0 & -1 \end{bmatrix} - \frac{q}{2N}\begin{bmatrix} 1 & 0 & 0 & 0 \\ 0 & 1 & 0 & 0 \\ 0 & 0 & -1 & 0 \\ 0 & 0 & 0 & -1 \end{bmatrix} \qquad \text{(II.4)}$$

We recognize in the first term the angular momentum component I_{zA} and, in the second term, I_{zX} [see (C13) and (C15)]. We note that the signs of the magnetic quantum numbers are in the same order as in Figure I.1 of Part I. We can now write $D(0)$ in shorthand:

$$D(0) = -\frac{p}{N} I_{zA} - \frac{q}{N} I_{zX} \qquad \text{(II.5)}$$

3. DESCRIBING THE EFFECT OF A PULSE IN TERMS OF ANGULAR MOMENTA

The rotation operator for a 90° pulse along the x-axis on nucleus X is:

$$R_{90xX} = \frac{1}{\sqrt{2}}\begin{bmatrix} 1 & 0 & i & 0 \\ 0 & 1 & 0 & i \\ i & 0 & 1 & 0 \\ 0 & i & 0 & 1 \end{bmatrix} \qquad \text{(see I.9)}$$

Its reciprocal is:

$$R_{90xX}^{-1} = \frac{1}{\sqrt{2}}\begin{bmatrix} 1 & 0 & -i & 0 \\ 0 & 1 & 0 & -i \\ -i & 0 & 1 & 0 \\ 0 & -i & 0 & 1 \end{bmatrix} \qquad \text{(see I.10)}$$

We postmultiply $D(0)$ from (II.3) with R_{90xX} (R, for brevity).

$$D(0)R = \frac{1}{2N}\begin{bmatrix} -p-q & 0 & 0 & 0 \\ 0 & p-q & 0 & 0 \\ 0 & 0 & -p+q & 0 \\ 0 & 0 & 0 & p+q \end{bmatrix} \frac{1}{\sqrt{2}}\begin{bmatrix} 1 & 0 & i & 0 \\ 0 & 1 & 0 & i \\ i & 0 & 1 & 0 \\ 0 & i & 0 & 1 \end{bmatrix}$$

$$= \frac{1}{2\sqrt{2}N}\begin{bmatrix} -p-q & 0 & -ip-iq & 0 \\ 0 & p-q & 0 & ip-iq \\ -ip+iq & 0 & -p+q & 0 \\ 0 & ip+iq & 0 & p+q \end{bmatrix} \tag{II.6}$$

We now premultiply this result by R^{-1}

$D(1) = R^{-1}[D(0)\ R]$

$$= \frac{1}{4N}\begin{bmatrix} 1 & 0 & -i & 0 \\ 0 & 1 & 0 & -i \\ -i & 0 & 1 & 0 \\ 0 & -i & 0 & 1 \end{bmatrix}\begin{bmatrix} -p-q & 0 & -ip-iq & 0 \\ 0 & p-q & 0 & ip-iq \\ -ip+iq & 0 & -p+q & 0 \\ 0 & ip+iq & 0 & p+q \end{bmatrix}$$

$$= \frac{1}{4N}\begin{bmatrix} -2p & 0 & -2iq & 0 \\ 0 & 2p & 0 & -2iq \\ 2iq & 0 & -2p & 0 \\ 0 & 2iq & 0 & 2p \end{bmatrix} = \frac{1}{2N}\begin{bmatrix} -p & 0 & -iq & 0 \\ 0 & p & 0 & -iq \\ iq & 0 & -p & 0 \\ 0 & iq & 0 & p \end{bmatrix} \tag{II.7}$$

Separation of p and q yields

$$D(1) = \frac{-p}{2N}\begin{bmatrix} 1 & 0 & 0 & 0 \\ 0 & -1 & 0 & 0 \\ 0 & 0 & 1 & 0 \\ 0 & 0 & 0 & -1 \end{bmatrix} + \frac{q}{2N}\begin{bmatrix} 0 & 0 & -i & 0 \\ 0 & 0 & 0 & -i \\ i & 0 & 0 & 0 \\ 0 & i & 0 & 0 \end{bmatrix} \tag{II.8}$$

Comparing the result with (C13) and (C14) we recognize I_{zA} and I_{yX} and we can write

$$D(1) = \frac{-p}{N} I_{zA} + \frac{q}{N} I_{yX} \tag{II.9}$$

Relations (II.5) and (II.9) open the way toward the product operator formalism. We succeeded in writing $D(0)$ and $D(1)$ in angular momentum shorthand. Moreover, we foresee the possibility of obtaining the result of pulses without matrix calculations; this can be seen by simply representing the angular momenta in their vector form, as in Figure II.1.

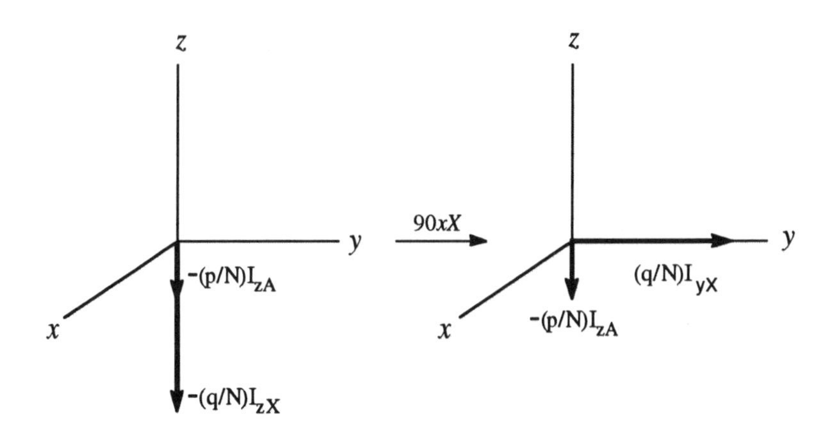

$$-(p/N)I_{zA} - (q/N)I_{zX} \xrightarrow{90xX} -(p/N)I_{zA} + (q/N)I_{yX}$$

Figure II.1. Effect of the 90*xX* pulse.

We note that, because of our convention to take gamma as negative (see Appendix J) the angular momentum vector orientation is opposite to that of the magnetization vector which was used in Part I.

4. AN UNSUCCESSFUL ATTEMPT TO DESCRIBE A COUPLED EVOLUTION IN TERMS OF ANGULAR MOMENTA

Let us calculate the result of a coupled evolution of duration $t_e/2$ starting from $D(1)$. Applying (I.13) to $D(1)$ in (II.7) gives:

$$D(2)=\frac{1}{2N}\begin{bmatrix} -p & 0 & F & 0 \\ 0 & p & 0 & G \\ F^* & 0 & -p & 0 \\ 0 & G^* & 0 & p \end{bmatrix} \tag{II.10}$$

where

$$\begin{aligned} F &= -iq\exp(-i\Omega_{13}t_e/2) \\ G &= -iq\exp(-i\Omega_{24}t_e/2) \end{aligned} \tag{II.11}$$

With the notations

$$\begin{aligned} \Omega_{13} &= \Omega_X + \pi J \\ \Omega_{24} &= \Omega_X - \pi J \end{aligned} \tag{II.12}$$

the exponentials in (II.11) become

$$\begin{aligned} \exp(-i\Omega_{13}t_e/2) &= \exp[-i(\Omega_X + \pi J)t_e/2] \\ &= \exp(-i\Omega_X t_e/2)\exp(-i\pi J t_e/2) \\ \exp(-i\Omega_{24}t_e/2) &= \exp[-i(\Omega_X - \pi J)t_e/2] \\ &= \exp(-i\Omega_X t_e/2)\exp(+i\pi J t_e/2) \end{aligned} \tag{II.13}$$

We make now the following notations:

$$\begin{aligned} c &= \cos\Omega_X t_e/2 \qquad & s &= \sin\Omega_X t_e/2 \\ C &= \cos\pi J t_e/2 \qquad & S &= \sin\pi\, J t_e/2 \end{aligned} \tag{II.14}$$

Note that here c and s have different meanings than the ones assigned in Part I (I.31). To make sure, c and s (lower case) represent effects of chemical shift and C and S (upper case) represent effects of J-coupling. Now the exponentials in (II.11) can be written as:

$$\begin{aligned} \exp(-i\Omega_{13}t_e/2) &= (c-is)(C-iS) \\ \exp(-i\Omega_{24}t_e/2) &= (c-is)(C+iS) \end{aligned} \tag{II.15}$$

Then F and G become

$$
\begin{aligned}
F &= -iq(c-is)(C-iS) = -iq(cC-sS) - q(sC+cS) \\
G &= -iq(c-is)(C+iS) = -iq(cC+sS) - q(sC-cS) \qquad \text{(II.16)}
\end{aligned}
$$

and we can rewrite D(2) as

$$
\frac{1}{2N}\begin{bmatrix} -p & 0 & \begin{matrix}-iq(cC-sS)\\ -q(sC+cS)\end{matrix} & 0 \\ 0 & p & 0 & \begin{matrix}-iq(cC+sS)\\ -q(sC-cS)\end{matrix} \\ \begin{matrix}iq(cC-sS)\\ -q(sC+cS)\end{matrix} & 0 & -p & 0 \\ 0 & \begin{matrix}iq(cC+sS)\\ -q(sC-cS)\end{matrix} & 0 & p \end{bmatrix} \qquad \text{(II.17)}
$$

We separate $D(2)$ into five matrices containing the factors p, qcC, qsS, qsC, and qcS:

$$
D(2) = -\frac{p}{2N}\begin{bmatrix} 1 & 0 & 0 & 0 \\ 0 & -1 & 0 & 0 \\ 0 & 0 & 1 & 0 \\ 0 & 0 & 0 & -1 \end{bmatrix} + \frac{q}{2N}cC\begin{bmatrix} 0 & 0 & -i & 0 \\ 0 & 0 & 0 & -i \\ i & 0 & 0 & 0 \\ 0 & i & 0 & 0 \end{bmatrix}
$$

$$
-\frac{q}{2N}sS\begin{bmatrix} 0 & 0 & -i & 0 \\ 0 & 0 & 0 & i \\ i & 0 & 0 & 0 \\ 0 & -i & 0 & 0 \end{bmatrix} - \frac{q}{2N}sC\begin{bmatrix} 0 & 0 & 1 & 0 \\ 0 & 0 & 0 & 1 \\ 1 & 0 & 0 & 0 \\ 0 & 1 & 0 & 0 \end{bmatrix}
$$

$$
-\frac{q}{2N}cS\begin{bmatrix} 0 & 0 & 1 & 0 \\ 0 & 0 & 0 & -1 \\ 1 & 0 & 0 & 0 \\ 0 & -1 & 0 & 0 \end{bmatrix} \qquad \text{(II.18)}
$$

In (II.18) only three terms, the first, second, and fourth, contain angular momenta [cf.(C12-C15)]. The first term is $-(p/N)I_{zA}$, the second is $(q/N)cCI_{yX}$ and the fourth is $-(q/N)sCI_{xX}$. The third and fifth matrices in (II.18) contain neither angular momenta nor a linear combination of the six components (I_{xA}, I_{yA}, I_{zA}, I_{xX}, I_{yX}, I_{zX}) known to us. This shows that the six angular momentum components shown in parenthesis are not sufficient to express the density matrix after a coupled evolution. In other words they constitute an *incomplete set* of *operators*. We will see in the next section how we can put together a *complete (basis) set* which will allow us to treat coupled evolutions in a shorthand similar to that used for $D(0)$ and $D(1)$. We may use as an analogy the blocks a child would need to build any number of castles of different shapes given in a catalog. For a given castle the child may not need to use all the building blocks, but he knows that none of the castles would require a block he does not have.

5. PRODUCT OPERATORS (PO)

We will call each building block a *basis operator* and will give in Table II.1 a complete set of such operators for the AX system. The bracket notations proposed by us will be defined as we explain how this set was put together.

Table II.1 Basis Operators for 2 Nuclei ($I = 1/2$)

$$[11] = \begin{bmatrix} 1 & 0 & 0 & 0 \\ 0 & 1 & 0 & 0 \\ 0 & 0 & 1 & 0 \\ 0 & 0 & 0 & 1 \end{bmatrix} \qquad [z1] = \begin{bmatrix} 1 & 0 & 0 & 0 \\ 0 & -1 & 0 & 0 \\ 0 & 0 & 1 & 0 \\ 0 & 0 & 0 & -1 \end{bmatrix}$$

$$[1z] = \begin{bmatrix} 1 & 0 & 0 & 0 \\ 0 & 1 & 0 & 0 \\ 0 & 0 & -1 & 0 \\ 0 & 0 & 0 & -1 \end{bmatrix} \qquad [zz] = \begin{bmatrix} 1 & 0 & 0 & 0 \\ 0 & -1 & 0 & 0 \\ 0 & 0 & -1 & 0 \\ 0 & 0 & 0 & 1 \end{bmatrix}$$

$$[x1]=\begin{bmatrix} 0 & 1 & 0 & 0 \\ 1 & 0 & 0 & 0 \\ 0 & 0 & 0 & 1 \\ 0 & 0 & 1 & 0 \end{bmatrix} \qquad [y1]=\begin{bmatrix} 0 & -i & 0 & 0 \\ i & 0 & 0 & 0 \\ 0 & 0 & 0 & -i \\ 0 & 0 & i & 0 \end{bmatrix}$$

$$[xz]=\begin{bmatrix} 0 & 1 & 0 & 0 \\ 1 & 0 & 0 & 0 \\ 0 & 0 & 0 & -1 \\ 0 & 0 & -1 & 0 \end{bmatrix} \qquad [yz]=\begin{bmatrix} 0 & -i & 0 & 0 \\ i & 0 & 0 & 0 \\ 0 & 0 & 0 & i \\ 0 & 0 & -i & 0 \end{bmatrix}$$

$$[1x]=\begin{bmatrix} 0 & 0 & 1 & 0 \\ 0 & 0 & 0 & 1 \\ 1 & 0 & 0 & 0 \\ 0 & 1 & 0 & 0 \end{bmatrix} \qquad [1y]=\begin{bmatrix} 0 & 0 & -i & 0 \\ 0 & 0 & 0 & -i \\ i & 0 & 0 & 0 \\ 0 & i & 0 & 0 \end{bmatrix}$$

$$[zx]=\begin{bmatrix} 0 & 0 & 1 & 0 \\ 0 & 0 & 0 & -1 \\ 1 & 0 & 0 & 0 \\ 0 & -1 & 0 & 0 \end{bmatrix} \qquad [zy]=\begin{bmatrix} 0 & 0 & -i & 0 \\ 0 & 0 & 0 & i \\ i & 0 & 0 & 0 \\ 0 & -i & 0 & 0 \end{bmatrix}$$

$$[xx]=\begin{bmatrix} 0 & 0 & 0 & 1 \\ 0 & 0 & 1 & 0 \\ 0 & 1 & 0 & 0 \\ 1 & 0 & 0 & 0 \end{bmatrix} \qquad [xy]=\begin{bmatrix} 0 & 0 & 0 & -i \\ 0 & 0 & -i & 0 \\ 0 & i & 0 & 0 \\ i & 0 & 0 & 0 \end{bmatrix}$$

$$[yx]=\begin{bmatrix} 0 & 0 & 0 & -i \\ 0 & 0 & i & 0 \\ 0 & -i & 0 & 0 \\ i & 0 & 0 & 0 \end{bmatrix} \qquad [yy]=\begin{bmatrix} 0 & 0 & 0 & -1 \\ 0 & 0 & 1 & 0 \\ 0 & 1 & 0 & 0 \\ -1 & 0 & 0 & 0 \end{bmatrix}$$

These product operators have been introduced by Ernst and coworkers (see O.W.Sörensen, G.W.Eich, M.H.Levitt, G.Bodenhausen, and R.R.Ernst in *Progress in NMR Spectroscopy*, **16**, 1983, 163-192, and references cited therein). The operators we use are multiplied by a factor of 2 in order to avoid writing 1/2 so many times.

Each basis operator is a product of two factors, one for each of the two nuclei (hence, the name "product operator"). The factor corresponding to a given nucleus may be one of its own angular momentum components multiplied by two ($2I_x$, $2I_y$, $2I_z$) or the unit matrix.

For example in the product operator [zz] the first factor is $2I_{zA}$ and the second, $2I_{zX}$. Proof:

$$\begin{bmatrix} 1 & 0 & 0 & 0 \\ 0 & -1 & 0 & 0 \\ 0 & 0 & 1 & 0 \\ 0 & 0 & 0 & -1 \end{bmatrix} \times \begin{bmatrix} 1 & 0 & 0 & 0 \\ 0 & 1 & 0 & 0 \\ 0 & 0 & -1 & 0 \\ 0 & 0 & 0 & -1 \end{bmatrix} = \begin{bmatrix} 1 & 0 & 0 & 0 \\ 0 & -1 & 0 & 0 \\ 0 & 0 & -1 & 0 \\ 0 & 0 & 0 & 1 \end{bmatrix}$$

$$2I_{zA} \qquad \times \qquad 2I_{zX} \qquad = \qquad [zz] \qquad \text{(II.19)}$$

Another example:

$$[x1] = 2I_{xA} \cdot [1] = 2I_{xA} = \begin{bmatrix} 0 & 1 & 0 & 0 \\ 1 & 0 & 0 & 0 \\ 0 & 0 & 0 & 1 \\ 0 & 0 & 1 & 0 \end{bmatrix} \qquad \text{(II.20)}$$

The basis set in Table II.1 allows us now to write (II.18) in shorthand because we recognize that the third term in (II.18) contains the product operator [zy] and the fifth term contains [zx]. Thus,

$$D(2) = -(p2N)[z1] + (q/2N)cC[1y] - (q/2N)sS[zy] \\ - (q/2N)sC[1x] - (q/2N)cS[zx] \qquad \text{(II.21)}$$

If we were to continue now to transform into product operators (PO) all the subsequent density matrices of HETCOR we would, of course, be able to do it, but this would do us no good. The real advantage will consist in finding a way to go from one PO to the next PO without using matrices. There is a small price to pay for this advantage, namely learning a few rules which show how to obtain a new PO representation after pulses or evolutions. It will be seen later that the PO advantage is much more important when we have to handle systems of more than two spins.

6. PULSE EFFECTS (ROTATIONS) IN THE PRODUCT OPERATOR FORMALISM

The great advantage of expressing the density matrix in terms of product operators is found in the dramatic simplification of calculations needed to describe pulse effects (rotations). Let us illustrate this by a few examples.

(1) 90*xX* pulse applied to *D*(0):

$$-(p/2N)[z1]-(q/2N)[1z]\xrightarrow{90xX}-(p/2N)[z1]+(q/2N)[1y]$$

_D_(0) **_D_(1)**

This PO operation can be readily visualized in vector representation. Indeed, looking at Figure II.2a we see that, while the angular momentum of X moves from $-z$ to $+y$ (90° rotation), the vector of A remains unaffected.

Let the vector representation guide us now to write another PO operation (see Figure II.2b).

(2) 90*xA* pulse applied to *D*(0):

$$-(p/2N)[z1]-(q/2N)[1z]\xrightarrow{90xA}+(p/2N)[y1]-(q/2N)[1z]$$

_D_(0) **_D_(1)**

By following the same procedure in examples 3 to 5 we find out that the product operators after any rotation can be written by changing the labels *x,y,z,* of the affected nucleus according to the rotation which took place in the vector representation. Of course, the unit matrix (label "1") always stays the same.

(3) 90*yAX* (nonselective) pulse applied to *D*(0):

$$-(p/2N)[z1]-(q/2N)[1z]\xrightarrow{90yAX}-(p/2N)[x1]-(q/2N)[1x]$$

D(0) **D(1)**

(4) 90xA applied to D(1) above:

$$-(p/2N)[x1]-(q/2N)[1x]\xrightarrow{90xA}-(p/2N)[x1]-(q/2N)[1x]$$

(No change, whatsoever)

(5) 90*yA* applied to *D*(1) above:

$$-(p/2N)[x1)-(q/2N)[1x]\xrightarrow{90yA}+(p/2N)[z1]-(q/2N)[1x]$$

The validity of this approach is demonstrated in Appendix E.

Many pulse sequences contain rotations other than 90° or 180°. We now proceed to apply our vector rule for an arbitrary angle α (see Figure II.2c). The PO representation of this rotation is

$$-(p/2N)[z1]-(q/2N)[1z]\xrightarrow{\alpha X}-(p/2N)[z1] \\ -(q/2N)([1z]\cos\alpha-[1y]\sin\alpha)$$

A few more examples of rotations are given below. This time we ignore the factors p/2N or q/2N.

$$[zz]\xrightarrow{90yA}[xz]$$

$$[zz]\xrightarrow{90yX}[zx]$$

$$[zz]\xrightarrow{90xA}-[yz]$$

$$[xy]\xrightarrow{180yA}-[xy]$$

$$[1y]\xrightarrow{90xA}[1y]$$

$$[zz]\xrightarrow{\alpha yA}[zz]\cos\alpha+[xz]\sin\alpha$$

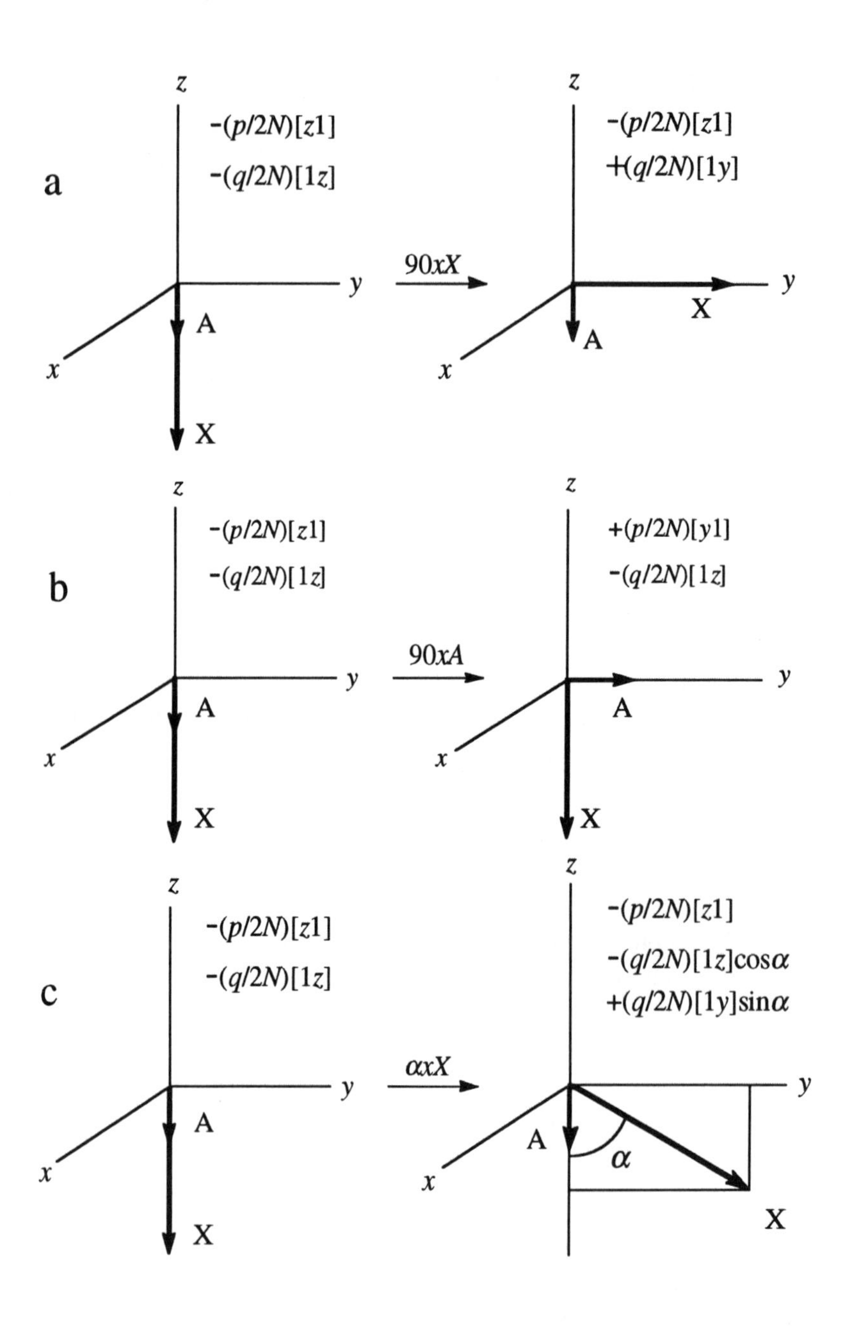

Figure II.2. PO and vector representation of rotations.

It must be mentioned at this time that only POs containing x, y, or z for one spin and 1 for the other(s) represent magnetization components along the corresponding Cartesian axes and have a vector representation. Nevertheless, the rotation applied to all other POs can be treated in the same way as above, by considering separately each factor in the product.

The great advantage over the density matrix formalism is that we can apply this approach to systems larger than two spins without the considerable increase in the computation volume (for CH_3 the matrix is 16x16, i.e., it has 256 elements). In an AMX system (three spin 1/2 nuclei):

$$\underset{AMX}{[zzz]} \xrightarrow{90yA} [xzz]$$

$$[xzz] \xrightarrow{180xM} -[xzz]$$

$$-[xzz] \xrightarrow{180xA} -[xzz]$$

Nonselective pulses affect more than one nucleus in the system. For example:

$$[zz] \xrightarrow{90xAX} [yy]$$

$$[xz] \xrightarrow{90xAX} -[xy]$$

$$[zzz] \xrightarrow{90yAM} [xxz]$$

For rotations α other than 90° or 180°, nonselective pulses affecting n nuclei must be handled in n successive operations. For instance, a nonselective pulse αxAX applied to the product operator $[zy]$ is treated in the following sequence:

$$[zy] \xrightarrow{\alpha xA} [zy]\cos\alpha - [yy]\sin\alpha$$

$$\xrightarrow{\alpha xX} [zy]\cos^2\alpha + [zz]\cos\alpha\sin\alpha - [yy]\sin\alpha\cos\alpha - [yz]\sin^2\alpha$$

These two operations may be performed in any order.

7. TREATMENT OF EVOLUTIONS IN THE PRODUCT OPERATOR FORMALISM

As shown in Appendix F, the evolution of coupled spins is conveniently treated in two steps. Step 1: we consider the system noncoupled (chemical shift evolution only). Step 2: we calculate the effect of coupling.

Step 1 (shift evolution)

A shift evolution is equivalent to a rotation about the z axis by an angle $\alpha = \Omega t$. Example:

$$[x1] \xrightarrow{\text{shift A}} [x1]\cos\Omega_A t + [y1]\sin\Omega_A t$$

The analogy with the vector representation is shown in Fig. II.3.

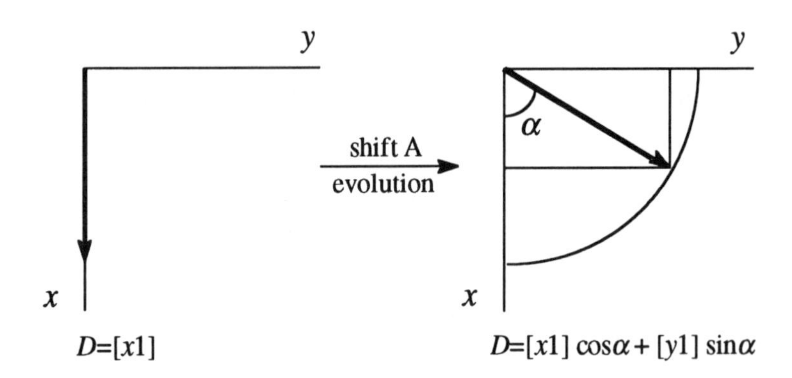

Figure II.3. PO and vector representation of a coupled evolution. The angle $\alpha = \Omega_A t$.

Another example:

$$[zy] \xrightarrow{\text{shift X}} [zy]\cos\Omega_X t - [zx]\sin\Omega_X t$$

The rule of thumb for shift evolution is :

PO after evolution = (PO before evolution) $\cos\Omega t$ + (PO before evolution in which x is replaced by y and y by $-x$ for the spin affected by evolution) $\sin\Omega t$. The labels 1 and z are invariant.

If more than one nucleus in the system is subject to shift evolution, these evolutions have to be treated as separate steps (the order is immaterial). Example:

$$\underset{\text{AMX}}{[xyz]} \xrightarrow{\text{shift A}} [xyz]\cos\Omega_A t + [yyz]\sin\Omega_A t$$

$$\begin{aligned} \xrightarrow{\text{shift M}} & [xyz]\cos\Omega_A t\cos\Omega_M t - [xxz]\cos\Omega_A t\sin\Omega_M t \\ & + [yyz]\sin\Omega_A t\cos\Omega_M t - [yxz]\sin\Omega_A t\sin\Omega_M t \end{aligned} \qquad \text{(II.22)}$$

The X spin is represented by z in the POs. Therefore the "shift X" does not bring any further change in (II.22).

With the notations

$$\begin{array}{lll} \cos\Omega_A t = c & \cos\Omega_M t = c' & \cos\Omega_X t = c'' \\ \sin\Omega_A t = s & \sin\Omega_M t = s' & \sin\Omega_X t = s'' \end{array} \qquad \text{(II.23)}$$

the relation (II.22) becomes

$$[xyz] \xrightarrow{\text{shift A, M, X}} cc'\,[xyz] - cs'\,[xxz] + sc'\,[yyz] - ss'\,[yxz]$$

One more example :

$$[xyy] \xrightarrow{\text{shift A}} c[xyy] + s[yyy]$$

$$\xrightarrow{\text{shift M}} cc'\,[xyy] - cs'\,[xxy] + sc'\,[yyy] - ss'\,[yxy]$$

$$\begin{aligned} \xrightarrow{\text{shift X}} & cc'\,c''[xyy] - cc'\,s''[xyx] - cs'\,c''[xxy] + cs'\,s''[xxx] \\ & + sc'\,c''[yyy] - sc'\,s''[yyx] - ss'\,c''[yxy] + ss'\,s''[yxx] \end{aligned}$$

Step 2 (J coupling evolution)

According to the rules presented in Appendix F the coupling between two spins is active only when one of the nuclei appears in the PO with an x or y while the other nucleus is represented by z or 1. Examples :

$\underset{\text{AMX}}{[xzz]}$ – couplings J_{AM} and J_{AX} are active, but J_{MX} is not.

$[xyz]$ – couplings J_{AX} and J_{MX} are active, but J_{AM} is not.

Another example:

[xyy] – no coupling is active[1]

For every coupling that is active, the step 2 rule is:
PO after evolution = (PO before evolution) $\cos\pi Jt$ + (PO before evolution with x replaced by y, y by $-x$, 1 by z,and z by 1) $\sin\pi Jt$.

Examples:

$$[xz] \xrightarrow{J_{AX}} C[xz] + S[y1] \qquad \text{with} \qquad \begin{matrix} C = \cos\pi J_{AX}t \\ S = \sin\pi J_{AX}t \end{matrix}$$

$$[xyz] \xrightarrow{J_{AX}} C[xyz] + S[yy1]$$

$$\xrightarrow{J_{MX}} CC'[xyz] - CS'[xx1] + SC'[yy1] - SS'[yxz]$$

$$\text{where} \qquad \begin{matrix} C = \cos\pi J_{AX}t \\ S = \sin\pi J_{AX}t \end{matrix} \qquad \begin{matrix} C' = \cos\pi J_{AM}t \\ S' = \sin\pi J_{AM}t \end{matrix}$$

Combinations of steps 1 and 2 are illustrated by two examples.

$$\text{System AM} \qquad [x1] \xrightarrow{\text{shift A}} c[x1] + s[y1]$$

$$\xrightarrow{J_{AM}} cC[x1] + cS[yz] + sC[y1] - sS[xz] \tag{II.24}$$

$$\text{System AMX} \qquad [x11] \xrightarrow{\text{shift A}} c[x11] + s[y11]$$

$$\xrightarrow{J_{AM}} cC[x11] + cS[yz1] + sC[y11] - sS[xz1]$$

$$\xrightarrow{J_{AX}} cCC'[x11] + cCS'[y1z] + cSC'[yz1] - cSS'[xzz]$$

$$+sCC'[y11] - sCS'[x1z] - sSC'[xz1] - sSS'[yzz] \tag{II.25}$$

Notations for the last two examples:

$$\begin{matrix} c = \cos\Omega_A t \\ s = \sin\Omega_A t \end{matrix} \qquad \begin{matrix} C = \cos J_{AM}t \\ S = \sin J_{AM}t \end{matrix} \qquad \begin{matrix} C' = \cos J_{AX}t \\ S' = \sin J_{AX}t \end{matrix} \tag{II.26}$$

We notice that, in the case of the two spin system, a coupled evolution does not split any PO into more than 4 terms. This is due to the fact that when both shifts are active the coupling is not.

---

[1] This seemingly surprising situation prompts an explanation. Product operators as [xx], [xy], [xyy] have nonvanishing elements on the secondary diagonal only (which represents zero- and multiple-quantum coherences). Referring to Figure I.1 we see that the evolution frequency for the double-quantum transition (1 → 4) is $(A+J/2)+(X-J/2) = X+A$. The zero-quantum transition frequency (2 →3) is X-A. None of them depends on J.

8. REFOCUSING ROUTINES

We have seen in the 2DHETCOR section that the $180xC$ pulse caused the decoupling of carbon from proton. In other words, the pulse applied in the middle of the evolution time t_e, caused the second half to compensate for the coupling effect of the first half. We call this a "refocusing routine." The chemical shift evolution can also be refocused if the pulse is applied on the nucleus that evolves. The routine, as shown below, can be handled in the conventional way (evolution-pulse-evolution) but the partial results are fairly more complicated than the final result.

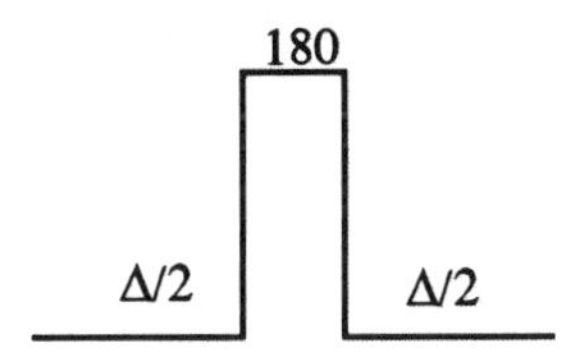

We suggest here an efficient calculation shortcut in which the entire evolution time, Δ, can be placed either before or after the pulse (of course, this cannot be done in the actual sequence).

During the hypothetical delay, Δ, the following rules apply:

a) Only shifts of the nuclei *not* affected by the 180° pulse are taken into account in the evolution Δ since all other are refocused.

b) The coupling between two nuclei is active if both or none of them are affected by the 180° pulse.

The above rules are valid for systems of *m* spin 1/2 nuclei, part of which may be magnetically equivalent. The 180° rotation is supposed to occur about an axis in the *xy* plane (no off-resonance pulse). The phase of the pulse does not affect the validity of the rules but it must be conserved when, in our calculations, we move the pulse from the middle of the interval Δ to one of its ends. The rules above are demonstrated in Appendix H.

We now compare the conventional calculation with the shortcut. For example, we assume that the density matrix at time $t(n)$ is

$$D(n) = p'[x1] + q'[1x]$$

and it is followed by

$$\Delta/2 - 180xA - \Delta/2$$

In the conventional way, i.e., evolution – pulse – evolution, we start with the first evolution, $\Delta/2$.

$$D(n) \xrightarrow{shift\ A} p'c[x1] + p's[y1] + q'[1x]$$
$$\xrightarrow{shift\ X} p'c[x1] + p's[y1] + q'c'[1x] + q's'[1y]$$
$$\xrightarrow{J} p'cC[x1] + p'cS[yz] + p'sC[y1] - p'sS[xz]$$
$$+q'c'C[1x] + q'c'S[zy] + q's'C[1y] - q's'S[zx] = D(n+1)$$

where

$$c = \cos\Omega_A\Delta/2 \qquad c' = \cos\Omega_X\Delta/2 \qquad C = \cos\pi J\Delta/2$$
$$s = \sin\Omega_A\Delta/2 \qquad s' = \sin\Omega_X\Delta/2 \qquad S = \sin\pi J\Delta/2$$

The 180xA pulse affects only the first label in the POs (nucleus A), changing the sign of y and z and leaving x unchanged.

$$D(n+1) \xrightarrow{180xA} p'cC[x1] - p'cS[yz] - p'sC[y1] - p'sS[xz]$$
$$+q'c'C[1x] - q'c'S[zy] + q's'C[1y] + q's'S[zx] = D(n+2)$$

The second evolution $\Delta/2$ is calculated as follows

$$D(n+2) \xrightarrow{shift\ A} p'c^2C[x1] + p'csC[y1] - p'c^2S[yz] + p'csS[xz]$$
$$-p'scC[y1] + p's^2C[x1] - p'scS[xz] - p's^2S[yz]$$
$$+q'c'C[1x] - q'c'S[zy] + q's'C[1y] + q's'S[zx]$$
$$\xrightarrow{shift\ X} p'c^2C[x1] + p'csC[y1] - p'c^2S[yz] + p'csS[xz]$$
$$-p'scC[y1] + p's^2C[x1] - p'scS[xz] - p's^2S[yz]$$
$$+q'c'^2C[1x] + q'c's'C[1y] - q'c'^2S[zy] + q'c's'S[zx]$$
$$+q's'c'C[1y] - q's'^2C[1x] + q's'c'S[zx] + q's'^2S[zy] = D(int)$$

This is an intermediate result, since we still have to consider the effect of J-coupling. Before doing so, we combine the terms containing the same PO.

$$D(int) = p'(c^2 + s^2)C[x1] - p'(c^2 + s^2)S[yz]$$
$$+q'(c'^2 - s'^2)C[1x] + 2q'c's'C[1y]$$
$$-q'(c'^2 - s'^2)S[zy] + 2q'c's'S[zx]$$

We recognize the expressions for the sine and cosine of twice the angle $\Omega_X\Delta/2$, i.e., $\Omega_X\Delta$.

$$D(int) = p'C[x1] - p'S[yz]$$
$$+q'\cos\Omega_X\Delta(C[1x] - S[zy]) + q'\sin\Omega_X\Delta(C[1y] + S[zx])$$

Further, we calculate the effect of J.

$$D(int) \xrightarrow{J} p'C^2[x1] + p'CS[yz] - p'SC[yz] + p'S^2[x1]$$
$$+q'\cos\Omega_X\Delta(C^2[1x] + CS[zy] - SC[zy] + S^2[1x])$$
$$+q'\sin\Omega_X\Delta(C^2[1y] - CS[zx] + SC[zx] + S^2[1y])$$
$$= p'[x1] + q'\cos\Omega_X\Delta[1x] + q'\sin\Omega_X\Delta[1y]$$

We now show that this result can be obtained in just two lines by using the shortcut. Following the rules described above, we first apply the 180*xA* pulse, then an evolution Δ (where only the shift X is considered, while the shift A and the coupling J are ignored).

$$D(n) \xrightarrow{180xA} p'[x1] + q'[1x]$$
$$\xrightarrow{shift\ X} p'[x1] + q'\cos\Omega_X\Delta[1x] + q'\sin\Omega_X\Delta[1y]$$

The first term, representing the A magnetization, appears unchanged because its evolution during the first delay $\Delta/2$ has been undone during the second $\Delta/2$. The X magnetization has evolved with the frequency Ω_X during the delay Δ, but in the end it is not affected by the J-coupling. This is because its two components, fast and slow, have undergone a change of label in the middle of the delay Δ.

In a system of two nuclei (A and X) if the pulse affects nucleus A, only shift X is operative. Shift A and coupling J_{AX} are refocused (see example above). If the 180° pulse affects both A and X, only the coupling is operative. Both shifts are refocused. In a system of more than two nuclei (e.g., AMX), if the pulse affects all nuclei, only the couplings are active. If the pulse affects all nuclei except one (e.g., nucleus A), we see the effect of shift A and of the couplings which do not involve A, i.e., J_{MX}. If the pulse affects only nucleus A, all shifts except A and the couplings involving A (J_{AM}, J_{AX}) are active.

9. PO TREATMENT OF 2DHETCOR: TWO SPINS (CH)

We consider again the sequence in Figure I.2 applied to a system of two weakly coupled spin 1/2 nuclei A and X. The density matrix at equilibrium (see II.5) is:

$$D(0) = -p'[z1] - q'[1z]$$

where

$$p' = p/2N$$
$$q' = q/2N$$
$$N = \text{number of states} = 4$$

p and q have the same meaning as in (I.3) and (I.4).

$$D(0) \xrightarrow{90xX} -p'[z1] + q'[1y] \qquad \text{(II.27)}$$
$$\mathbf{D(1)}$$

We apply the "refocusing routine" treatment to the segment

$$t_e/2 - 180xA - t_e/2$$

As shown in Section II.8 this routine has the same effect as the $180xA$ pulse followed by an evolution t_e during which only the shift X is active. The coupling AX is refocused by the $180xA$ pulse.

$$D(1) \xrightarrow{180xA} p'[z1] + q'[1y]$$
$$\xrightarrow{t_e(shift\ X)} p'[z1] + q' \cos W_X t_e [1y] - q' \sin W_X t_e [1x] \qquad \text{(II.28)}$$
$$\mathbf{D(4)}$$

A coupled evolution Δ_1 follows, which can be handled according to the rules of Section II.7 with

$$c' = \cos \Omega_X \Delta_1 \qquad C = \cos \pi J \Delta_1$$
$$s' = \sin \Omega_X \Delta_1 \qquad S = \sin \pi J \Delta_1$$

$$D(4) \xrightarrow{\text{shift A}} \text{same}$$
$$\xrightarrow{\text{shift X}} p'[z1] + q' \cos \Omega_X t_e (c'[1y] - s'[1x]) - q' \sin \Omega_X t_e (c'[1x] + s'[1y])$$
$$\xrightarrow{J} p'[z1] + q' \cos \Omega_X t_e (c'\,C[1y] - c'\,S[zx] - s'\,C[1x] - s'\,S[zy])$$
$$-q' \sin \Omega_X t_e (c'\,C[1x] + c'\,S[zy] + s'\,C[1y] - s'\,S[zx])$$
$$\mathbf{D(5)}$$

Since $\Delta_1 = 1/2J$ and $\pi J\Delta_1 = \pi/2$, $C = 0$ and $S = 1$

$$D(5) = p'[z1] + q' \cos\Omega_X t_e(-c'[zx] - s'[zy])$$
$$-q' \sin\Omega_X t_e(-s'[zx] + c'[zy]) \qquad \text{(II.29)}$$

Using the trigonometric relations for the sum of two angles [see (A29) and (A30)] we can rewrite $D(5)$:

$$D(5) = p'[z1] + q'[zx](\cos\Omega_X t_e \cos\Omega_X\Delta_1 - \sin\Omega_X t_e \sin\Omega_X\Delta_1)$$
$$-q'[zy](\cos\Omega_X t_e \sin\Omega_X\Delta_1 + \sin\Omega_X t_e \cos\Omega_X\Delta_1)$$
$$= p'[z1] - q'[zx]\cos\Omega_X(t_e + \Delta_1) - q'[zy]\sin\Omega_X(t_e + \Delta_1) \qquad \text{(II.30)}$$

We calculate now the effects of the pulses three and four (in the PO formalism it is simpler to handle them separately):

$$D(5) \xrightarrow{90xX} p'[z1] - q'[zx]\cos\Omega_X(t_e + \Delta_1) - q'[zz]\sin\Omega_X(t_e + \Delta_1)$$
$$\mathbf{D(6)}$$
$$\xrightarrow{90xA} -p'[y1] + q'[yx]\cos\Omega_X(t_e + \Delta_1) + q'[yz]\sin\Omega_X(t_e + \Delta_1)$$
$$\mathbf{D(7)} \qquad \text{(II.31)}$$

Since no other r.f. pulse follows we can concentrate on those terms which represent observable magnetization components. We observe nucleus A, therefore we are interested in the product operators [x1] and [y1] which give M_{xA} and M_{yA}. We are also interested in [yz] and [xz] which can evolve into [x1], [y1] during a coupled evolution. All product operators other than the four mentioned above are nonobservable terms (NOT). We can rewrite $D(7)$ as:

$$D(7) = -p'[y1] + q'[yz]\sin\Omega_X(t_e + \Delta_1) + \text{NOT} \qquad \text{(II.32)}$$

The second term is important for the 2D experiment because it is proton modulated (it contains the frequency Ω_X). It is also enhanced by polarization transfer (i.e., multiplied by q' rather than p').

The coupled evolution Δ_2 is necessary to render the second term observable. If the decoupled detection started at $t(7)$, the [yz] term would evolve into a combination of [yz] and [xz], none of which is observable. The observable terms [x1] and [y1] can derive from [yz] only in a coupled evolution.

$$D(7) \xrightarrow{\text{shift A}} -p'(c[y1] - s[x1]) + q' \sin\Omega_X(t_e + \Delta_1)(c[yz] - s[xz])$$
$$\xrightarrow{\text{shift X}} \text{same} \xrightarrow{J} -p'(cC[y1] - cS[xz] - sC[x1] - sS[yz])$$
$$+q' \sin\Omega_X(t_e + \Delta_1)(cC[yz] - cS[x1] - sC[xz] - sS[y1]) + \text{NOT} \quad \text{(II.33)}$$

$D(8)$

where

$$c = \cos\Omega_A\Delta_2 \qquad C = \cos\pi J\Delta_2$$
$$s = \sin\Omega_A\Delta_2 \qquad S = \sin\pi J\Delta_2 \qquad \text{(II.34)}$$

During the decoupled evolution that follows after $t(8)$, the product operators [x1], [y1] will evolve into combinations of [x1], [y1] while the product operators [xz], [yz] evolve into combinations of [xz],[yz], i.e., they will remain nonobservable. We can therefore retain only [x1], [y1] in the explicit expression of $D(8)$:

$$D(8) = -p'(cC[y1] - sC[x1]) + q' \sin\Omega_X(t_e + \Delta_1)(-cS[x1] - sS[y1])$$
$$+ \text{NOT} \qquad \text{(II.35)}$$

In order to maximize the proton modulated term, one selects for Δ_2 the value $1/2J$ which leads to $S = 1$ and $C = 0$. This value of Δ_2 represents an optimum in the particular case of the AX system. It will be shown in Section II.10 that for AX_2 and AX_3 (e.g., the methylene and methyl cases) a shorter Δ_2 is to be used. For $\Delta_2 = 1/2J$:

$$D(8) = -q' \sin\Omega_X(t_e + \Delta_1)(\cos\Omega_A\Delta_2[x1] + \sin\Omega_A\Delta_2[y1]) + \text{NOT}$$
$$\text{(II.36)}$$

The simplest way to describe the decoupled evolution t_d (from the point of view of the observable A) is a rotation of the transverse magnetization M_{TA} about the z axis. At $t(8)$ we have (see Appendix J):

$$M_{xA}(8) = -(M_{oA}/p')(\text{coefficient of } [x1])$$

$$= (q'/p')M_{oA}\sin\Omega_X(t_e+\Delta_1)\cos\Omega_A\Delta_2$$

$$M_{yA}(8) = (q'/p')M_{oA}\sin\Omega_X(t_e+\Delta_1)\sin\Omega_A\Delta_2$$

$$M_{TA}(8) = M_{xA}(8)+iM_{yA}(8) = (q'/p')M_{oA}\sin\Omega_X(t_e+\Delta_1)\exp(i\Omega_A\Delta_2) \quad \text{(II.37)}$$

The ratio $q'/p' = \gamma_X/\gamma_A$ represents the enhancement factor through polarization transfer.

By handling the decoupled evolution as a magnetization rotation we get

$$\begin{aligned} M_{TA}(9) &= M_{TA}(8)\exp(i\Omega_A t_d) \\ &= (q'/p')M_{oA}\sin\Omega_X(t_e+\Delta_1)\exp[i\Omega_A(t_d+\Delta_2)] \end{aligned} \quad \text{(II.38)}$$

This is in agreement with the result obtained in Part I through DM calculations. The calculations requested by the PO approach are somewhat less complicated than those of the DM approach. The real advantage will be seen when we apply (see Section II.10) the PO formalism to an AX_2 and AX_3 case (e.g., the 2DHETCOR of a CH_2 or CH_3).

10. PO TREATMENT OF 2DHETCOR: CH_2 AND CH_3

We extend the calculations carried out in section II.9 to an AX_2 or AX_3 system (e.g., a methylene or a methyl). The density matrix at equilibrium is:

$$\begin{aligned} D(0) &= -p'[z11] - q'([1z1]+[11z]) && \text{for } AX_2 \\ D(0) &= -p'[z111] - q'([1z11]+[11z1]+[111z]) && \text{for } AX_3 \end{aligned} \quad \text{(II.39)}$$

Instead of following the two cases separately, we use the "multiplet formalism" introduced in Appendix L. The reader should get acquainted with this formalism before proceeding further.

$$D(0) = -(p'/n)\{z1\} - q'\{1z\} \quad \text{(II.40)}$$

valid for any AX_n system.

$$D(0) \xrightarrow{90xX} -(p'/n)\{z1\} + q'\{1y\} \quad \text{(II.41)}$$

D(1)

As we did in section II.9, we treat the segment $t_e/2 - 180xA - t_e/2$

as a "refocusing routine" and this brings us directly from $D(1)$ to $D(4)$.

$$D(1) \xrightarrow{180xA} +(p'/n)\{z1\} + q'\{1y\} \xrightarrow{t_e(\text{shiftX})} (p'/n)\{z1\} + q'\cos\Omega_X t_e\{1y\} - q'\sin\Omega_X t_e\{1x\} \tag{II.42}$$

D(4)

With the assumption that $\Delta_1 = 1/2J$ (i.e., $\pi J\Delta_1 = \pi/2$) we have

$$D(4) \xrightarrow{\Delta_1(J)} (p'/n)\{z1\} - q'\cos\Omega_X t_e\{zx\} - q'\sin\Omega_X t_e\{zy\}$$

$$\xrightarrow{\Delta_1(\text{shift X})} (p'/n)\{z1\} - q'\cos\Omega_X t_e(c'\{zx\} + s'\{zy\}) - q'\sin\Omega_X t_e(c'\{zy\} - s'\{zx\}) \tag{II.43}$$

D(5)

where c', s' have the same meaning as in (II.29), i.e.,

$$c' = \cos\Omega_X\Delta_1 \quad ; \quad s' = \sin\Omega_X\Delta_1$$

Using again the trigonometric relations for the sum of two angles as we did in (II.30) we rewrite $D(5)$ as:

$$D(5) = (p'/n)\{z1\} - q'\{zx\}\cos\Omega_X(t_e + \Delta_1) - q'\{zy\}\sin\Omega_X(t_e + \Delta_1) \tag{II.44}$$

$$D(5) \xrightarrow{90xX} (p'/n)\{z1\} - q'\{zx\}\cos\Omega_X(t_e + \Delta_1) - q'\{zz\}\sin\Omega_X(t_e + \Delta_1)$$

D(6)

$$D(6) \xrightarrow{90xA} -(p'/n)\{y1\} + q'\{yx\}\cos\Omega_X(t_e + \Delta_1) + q'\{yz\}\sin\Omega_X(t_e + \Delta_1)$$

D(7) (II.45)

$$D(7) = -(p'/n)\{y1\} + q'\{yz\}\sin\Omega_X(t_e + \Delta_1) + \text{NOT} \tag{II.46}$$

Up to this point we have followed step by step the calculations in Section II.9, while formally replacing [ ] by { } and using p'/n instead of p'. This procedure is always valid for rotations (pulses) and in this case it was allowed for evolutions as stated in Appendix L, rule #4.

For the coupled evolution Δ_2 we can concentrate on the observable terms $\{x1\}$, $\{y1\}$ only and take advantage of the exception stated in Appendix L, rule #5.

$$D(7) \xrightarrow{\Delta_2} -(p'/n)(-sC^n\{x1\} + cC^n\{y1\})$$
$$+q' \sin \Omega_X (t_e + \Delta_1)(-cSC^{n-1}\{x1\} - sSC^{n-1}\{y1\}) + \text{NOT} \qquad \text{(II.47)}$$

D(8)

n= number of magnetically equivalent X nuclei (e.g., number of protons in CH_n).

$c = \cos \Omega_A \Delta_2$ $\qquad$ $C = \cos \pi J \Delta_2$
$s = \sin \Omega_A \Delta_2$ $\qquad$ $S = \sin \pi J \Delta_2$

In order to separate the effect of shift (c and s) and coupling (C and S) we rewrite D(8) as

$$D(8) = -(p'/n)C^n(c\{y1\} - s\{x1\}) - q' \sin \Omega_X (t_e + \Delta_1) SC^{n-1}(c\{x1\} + s\{y1\}) \qquad \text{(II.48)}$$

Here again the second term is the true 2D signal while the first one generates axial peaks in the 2D picture since it does not contain the frequency Ω_X.

The value $\Delta_2 = 1/2J$, which implies $S = 1$ and $C = 0$, is not suitable anymore because it nulls the useful 2D signal for any n >1. We will discuss later the optimum value of Δ_2.

In writing the magnetization components we have to follow the procedure stated as rule #6 in Appendix L.

$$M_{xA}(8) = -(nM_{oA}/p')(\text{coefficient of } \{x1\})$$
$$= -M_{oA}C^n \sin \Omega_A \Delta_2 + nM_{oA}(q'/p')SC^{n-1} \sin \Omega_X (t_e + \Delta_1) \cos \Omega_A \Delta_2$$

$$M_{yA}(8) = -(nM_{oA}/p')(\text{coefficient of } \{y1\})$$
$$= M_{oA}C^n \cos \Omega_A \Delta_2 + nM_{oA}(q'/p')SC^{n-1} \sin \Omega_X (t_e + \Delta_1) \sin \Omega_A \Delta_2$$

$$M_{TA}(8) = M_{xA}(8) + iM_{yA}(8)$$

$$= iM_{oA}C^n \exp(i\Omega_A \Delta_2) + nM_{oA}(q'/p')SC^{n-1} \sin \Omega_X (t_e + \Delta_1) \exp(i\Omega_A \Delta_2)$$

$$= M_{oA}[iC^n + n(q'/p')SC^{n-1} \sin \Omega_X (t_e + \Delta_1)] \exp(i\Omega_A \Delta_2) \qquad \text{(II.49)}$$

The decoupled evolution t_d during which the detection takes place is treated the same way it was done in II.9, as a rotation about Oz.

$$M_{TA}(9) = M_{TA}(8)\exp(i\Omega_A t_d)$$
$$= M_{oA}[iC^n + n(q'/p')SC^{n-1}\sin\Omega_X(t_e+\Delta_1)]\exp[i\Omega_A(t_d+\Delta_2)] \quad \text{(II.50)}$$

The enhancement factor of the 2D term is

$$n(q'/p')SC^{n-1}.$$

Now we can discuss the optimum value of Δ_2. In Figure II.4 the value of the product nSC^{n-1} is plotted versus $\Delta_2 J$ for n = 1, 2 and 3. The optimum Δ_2 values are:

$$n=1\ (CH) \longrightarrow \Delta_2 = 0.5/J$$
$$n=2\ (CH_2) \longrightarrow \Delta_2 = 0.25/J \quad \text{(II.51)}$$
$$n=3\ (CH_3) \longrightarrow \Delta_2 = 0.196/J$$

A good compromise is Δ_2 = $0.3/J$ for which all three expressions S, $2SC$, and $3SC^2$ have values exceeding 0.8. The bad news is that this Δ_2 value does not cancel the axial peak, represented by the term iC^n in (II.50).

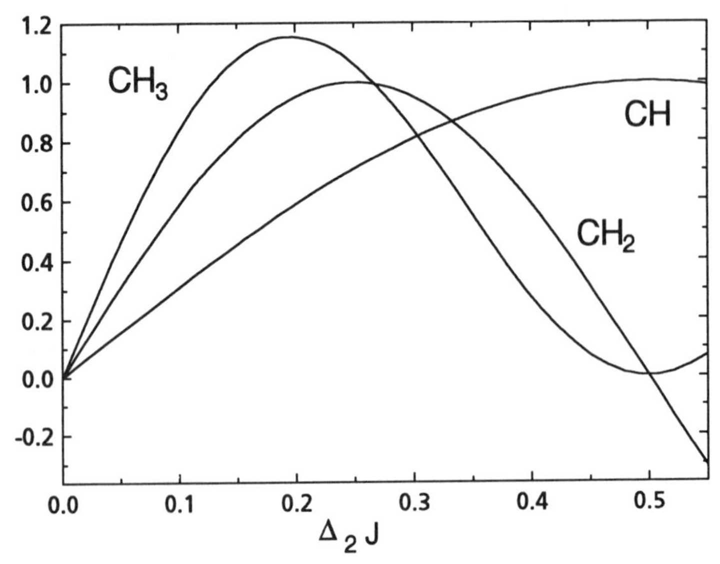

Figure II.4. Dependence of the factor nSC^{n-1} (S for CH, $2SC$ for CH_2, and $3SC^2$ for CH_3) on $\Delta_2 J$.

11. PO TREATMENT OF A POLARIZATION TRANSFER SEQUENCE: INEPT (INSENSITIVE NUCLEI ENHANCEMENT BY POLARIZATION TRANSFER) WITH DECOUPLING

We discuss first the decoupled INEPT sequence shown in Figure II.5. Its goal, an increased sensitivity of ^{13}C spectra, is achieved in two ways: by increasing the peak intensities for the protonated carbons, and by allowing a larger number of scans in a given experiment time.

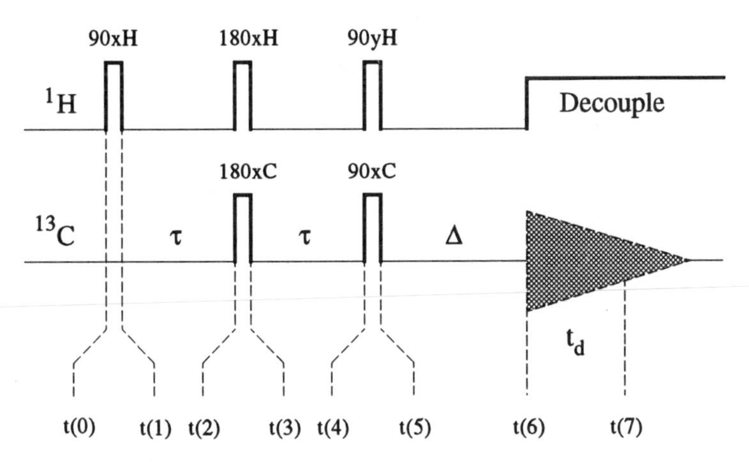

Figure II.5 The INEPT sequence: $90xH - \tau - 180xCH - \tau - 90yH - 90xC - \Delta - AT$

We treat the CH_n case with $n = 1,2$ or 3. The density matrix at thermal equilibrium is:

$$D = -(p'/n)\{z1\} - q'\{1z\} \tag{II.52}$$

One of the advantages of INEPT is that it allows fast scanning, (limited by the proton relaxation only). In order to emphasize this feature we will assume that only the protons have fully relaxed in the interval between sequences and will write the initial density matrix as

$$D(0) = -\lambda(p'/n)\{z1\} - q'\{1z\} \tag{II.53}$$

where λ is a recovery factor for carbon $(0 \leq \lambda \leq 1)$.

The effect of the first pulse is

$$D(0) \xrightarrow{90xH} -\lambda(p'/n)\{z1\} + q'\{1y\} \qquad \text{(II.54)}$$

D(1)

See Appendix L for the meaning of { } (multiplet formalism). We treat now the portion from $t(1)$ to $t(4)$ as a shift-refocusing routine (see Section II.8). We will apply a nonselective $180xCH$ pulse followed by a 2τ evolution in which the coupling only is expressed and this will bring us to $t(4)$.

$$D(1) \xrightarrow{180xCH} +\lambda(p'/n)\{z1\} - q'\{1y\} \xrightarrow{2\tau\ (J\ coupl.)} \qquad \text{(II.55)}$$
$$\longrightarrow \lambda(p'/n)\{z1\} - q'\{1y\}\cos 2\pi J\tau + q'\{zx\}\sin 2\pi J\tau$$

D(4)

The next two pulses have to be treated successively because one of the terms in $D(4)$ is affected by both the proton and the carbon pulse.

$$D(4) \xrightarrow{90yH} \lambda(p'/n)\{z1\} - q'\{1y\}\cos 2\pi J\tau - q'\{zz\}\sin 2\pi J\tau$$
$$\xrightarrow{90xC} -\lambda(p'/n)\{y1\} - q'\{1y\}\cos 2\pi J\tau + q'\{yz\}\sin 2\pi J\tau$$

D(5) (II.56)

After $t(5)$ no pulse follows and we can concentrate on the observable terms, keeping in mind that our observe nucleus is ^{13}C.

$$D(5) = -\lambda(p'/n)\{y1\} + q'\{yz\}\sin 2\pi J\tau + \text{NOT} \qquad \text{(II.57)}$$

The second term is enhanced by polarization transfer and it does not depend on λ. This means the pulse repetition rate is limited only by the proton relaxation as far as the second term is concerned. The optimum value for τ is $1/4J$ which leads to $\sin 2\pi J\tau = 1$. With this assumption:

$$D(5) = -\lambda(p'/n)\{y1\} + q'\{yz\} + \text{NOT} \qquad \text{(II.58)}$$

In $D(5)$ the second term is still not an observable; hence, the necessity of Δ. Using rule #5 in Appendix L we obtain:

$$D(5) \xrightarrow{\Delta} -\lambda(p'/n)C^n(c\{y1\} - s\{x1\})$$
$$+ q'SC^{n-1}(-c\{x1\} - s\{y1\}) + \text{NOT} \qquad \text{(II.59)}$$

D(6)

where

$$c = \cos \Omega_C \Delta \qquad C = \cos \pi J \Delta$$
$$s = \sin \Omega_C \Delta \qquad S = \sin \pi J \Delta$$

The expression (II.59) has much in common with (II.48), with the difference that the proton frequency Ω_H is not to be seen. This is not alarming since INEPT is not intended as a 2D sequence. Going through the same steps as from (II.48) to (II.50) we get:

$$M_{TC}(6) = M_{oC}[-i\lambda C^n - n(q'/p')SC^{n-1}]\exp(i\Omega_C \Delta) \qquad \text{(II.60)}$$

$$M_{TC}(7) = M_{oC}[-i\lambda C^n - n(q'/p')SC^{n-1}]\exp[i\Omega_C(t_d + \Delta)] \qquad \text{(II.61)}$$

We focus on the second term in the brackets (polarization transfer and fast scanning). The optimum value of Δ is selected according to Figure II.4 and relations (II.51), leading to $nSC^{n-1} \geq 0.8$. The enhancement factor is therefore $\geq 0.8\gamma_H / \gamma_C$.

12. COUPLED INEPT

The coupled INEPT (Figure II.6) is used for spectra editing.

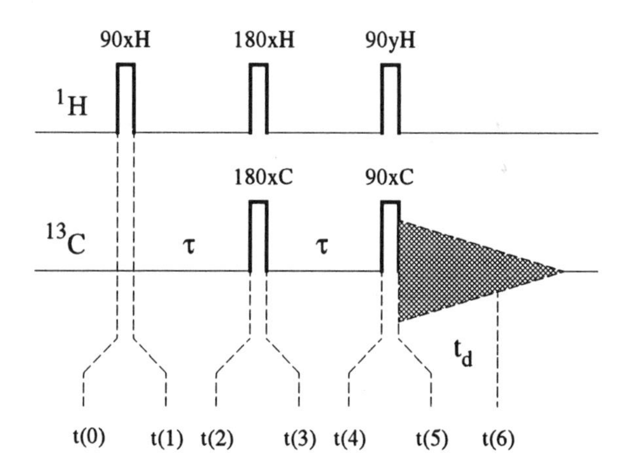

Figure II.6 The coupled INEPT sequence: $90xH - \tau - 180xCH - \tau - 90yH - 90xC - AT$

The magnetization at $t(6)$ can be written by taking its expression from the previous sequence (II.60) and replacing Δ by t_d:

$$M_{TC}(6) = M_{oC}[-i\lambda C^n - n(q'/p')SC^{n-1}]\exp(i\Omega_C t_d) \qquad \text{(II.62)}$$

with $$C = \cos\pi J t_d \qquad S = \sin\pi J t_d \qquad \text{(II.63)}$$

The resemblance with (II.61) is only formal. In (II.61) we had:

$$C = \cos\pi J\Delta \qquad S = \sin\pi J\Delta$$

and the detection time t_d appeared only in the exponential factor outside the brackets. The signal was a singlet. In (II.62) the variable t_d is also contained in C and S and the Fourier transform exhibits a multiplet. In order to see how this multiplet looks like, we have to discuss the expression (II.62) for $n = 1$, 2, and 3. We split $M_{TC}(6)$ into two terms

$$M_{TC}(6) = -M_{oC}(q'/p')nSC^{n-1}\exp(i\Omega_C t_d) - iM_{oC}\lambda C^n \exp(i\Omega_C t_d)$$
$$= M_A + M_B \qquad \text{(II.64)}$$

and discuss these two terms separately, while keeping in mind that C and S have the meanings in (II.63).

A) Polarization enhanced multiplet (term M_A)

We discuss separately the CH, CH_2, and CH_3 cases.

$n = 1$ (CH case)

$$M_A = -M_{oC}(q'/p')S\exp(i\Omega_C t_d)$$

Using (A28) we have

$$S = \frac{\exp(i\pi J t_d) - \exp(-i\pi J t_d)}{2i}$$

and this leads to

$$M_A = M_{oC}(q'/p')(\frac{1}{2i})[-e^{i(\Omega_C + \pi J)t_d} + e^{i(\Omega_C - \pi J)t_d}] \qquad \text{(II.65)}$$

This is the up-down doublet of Figure II.7a .

For a methylene

$n = 2$ (CH_2 case)

$$M_A = -M_{oC}(q'/p')2SC\exp(i\Omega_C t_d)$$

Using (A36) we have

$$SC = \frac{\exp(2i\pi Jt_d) - \exp(-2i\pi Jt_d)}{4i}$$

and

$$M_A = M_{oC}(q'/p')(\frac{1}{2i})[-e^{i(\Omega_C + 2\pi J)t_d} + e^{i(\Omega_C - 2\pi J)t_d}] \qquad \text{(II.66)}$$

This is a triplet with the central line missing and the other two lines one up and one down (see Figure II.7b).

$n = 3$ (CH_3 case)

$$M_A = -M_{oC}(q'/p')3SC^2\exp(i\Omega_C t_d)$$

We use (A38) and obtain

$$M_A = M_{oC}(q'/p')(\frac{3}{8i})[-e^{i(\Omega_C + 3\pi J)t_d} - e^{i(\Omega_C + \pi J)t_d} + e^{i(\Omega_C - \pi J)t_d} + e^{i(\Omega_C - 3\pi J)t_d}] \qquad \text{(II.67)}$$

This is the peculiar quartet in Figure II.7c: four lines of equal intensities, two up and two down.

B) Residual nonenhanced multiplet (term M_B)

The term M_B in (II.64) is smaller than M_A (no polarization enhancement and $\lambda < 1$). We will show that it represents a conventional multiplet.

$n = 1$ (CH case)

$$M_B = -iM_{oC}\lambda C\exp(i\Omega_C t_d)$$

Using (A27) and $-i = 1/i$ we obtain

$$M_B = M_{oC}\lambda(\frac{1}{2i})[e^{i(\Omega_C+\pi J)t_d} + e^{i(\Omega_C-\pi J)t_d}] \qquad (II.68)$$

the regular doublet in Figure II.7d .

n = 2 (CH_2 case)

$$M_B = -iM_{oC}\lambda C^2 \exp(i\Omega_C t_d) \quad .$$

Using (A35) we obtain the triplet

$$M_B = M_{oC}\lambda(\frac{1}{4i})[e^{i(\Omega_C+2\pi J)t_d} + 2e^{i\Omega_C t_d} + e^{i(\Omega_C-2\pi J)t_d}] \quad (II.69)$$

as in Figure II.7e .

$n = 3$ (CH_3 case)

$$M_B = -iM_{oC}\lambda C^3 \exp(i\Omega_C t_d)$$

We use (A37) and obtain

$$M_B = M_{oC}\lambda(\frac{1}{8i})[e^{i(\Omega_C+3\pi J)t_d} + 3e^{i(\Omega_C+\pi J)t_d} + 3e^{i(\Omega_C-\pi J)t_d} + e^{i(\Omega_C-3\pi J)t_d}]$$

$$(II.70)$$

a regular looking quartet (see Figure II.7f).

We notice that all the expressions (II.65) through (II.70) contain the factor $1/i$ (or $-i$) , indicating that the respective magnetizations are along the $-y$ axis.

When the term M_B is not vanishingly small, it breaks the symmetry of the multiplet, as shown in Figure II.7 (g through i). This drawback can be eliminated by means of an appropriate phase cycling. It is shown in Section II.13 that we have a similar situation with the DEPT sequence and a two step phase cycling is sufficient to cancel the residual nonenhanced signal.

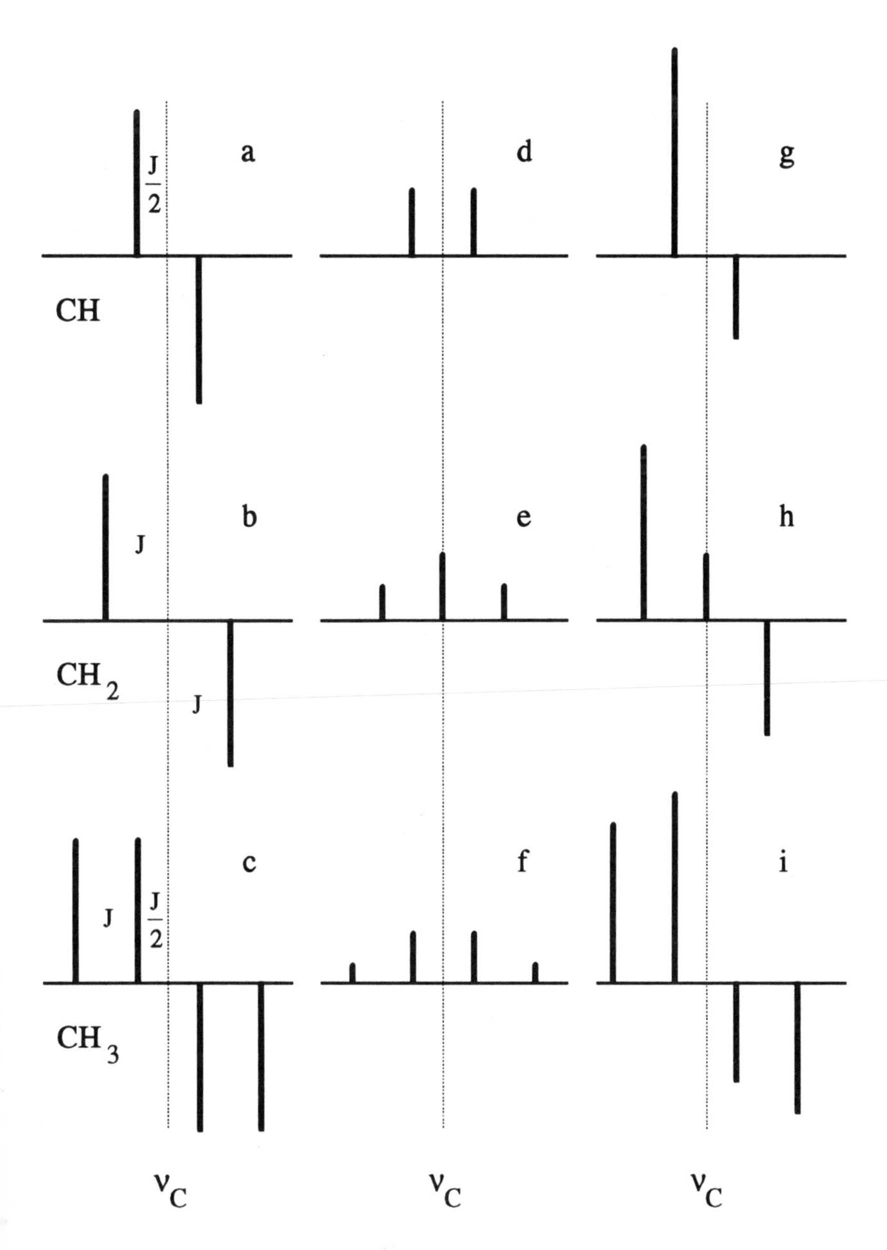

Figure II.7. Coupled INEPT; a,b,c – Polarization enhanced multiplet (term M_A); d,e,f – Residual non-enhanced multiplet (term M_B); g,h,i – Actual spectrum (term $M_A + M_B$)

13. PO TREATMENT OF DEPT (DISTORTIONLESS ENHANCEMENT POLARIZATION TRANSFER)

DEPT is a one-dimensional sequence used as a tool for unambiguous identification of the CH, CH_2, and CH_3 peaks in a proton decoupled ^{13}C spectrum. It shares with INEPT the advantage of permitting a fast repetition rate. The recycle time has to be longer than the proton relaxation time but can be fairly shorter than the carbon T_1. The nonprotonated carbons will not show up in a DEPT spectrum.

The sequence is shown in Figure II.8.

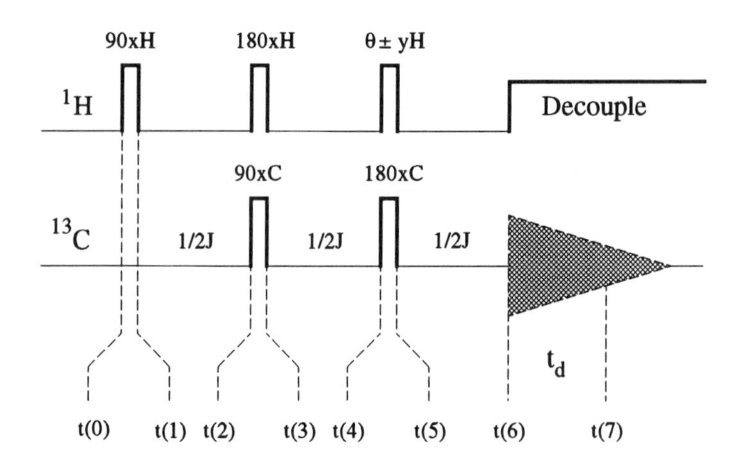

Figure II.8. The basic (one-dimensional) DEPT sequence: $90xH - 1/2J - 180xH - 90xC - 1/2J - 180xC - \theta \pm yH - 1/2J - AT(dec)$.

The initial density matrix is

$$D(0) = -\lambda(p'/n)\{z1\} - q'\{1z\} \tag{II.71}$$

where λ is a recovery factor for carbon [see (II.53)]

$$D(0) \xrightarrow{90xH} -\lambda(p'/n)\{z1\} + q'\{1y\} \tag{II.72}$$

$$D(1)$$

For the evolution we use the notations

$$c = \cos\Omega_C / 2J \qquad c' = \cos\Omega_H / 2J \qquad C = \cos\pi J / 2J = \cos\pi / 2 = 0$$
$$s = \sin\Omega_C / 2J \qquad s' = \sin\Omega_H / 2J \qquad S = \sin\pi J / 2J = \sin\pi / 2 = 1$$

(II.73)

In treating the evolution, the shift and coupling may be handled in any order. Here, because of the particular values of C and S, we are better off by starting with the coupling.

$$D(1) \xrightarrow{J\text{ coupl.}} -\lambda(p'/n)\{z1\} - q'\{zx\}$$

$$\xrightarrow{\text{shift H}} -\lambda(p'/n)\{z1\} - q'c'\{zx\} - q's'\{zy\} \qquad \text{(II.74)}$$

$$\boldsymbol{D(2)}$$

$$D(2) \xrightarrow{180xH} -\lambda(p'/n)\{z1\} - q'c'\{zx\} + q's'\{zy\}$$

$$\xrightarrow{90xC} \lambda(p'/n)\{y1\} + q'c'\{yx\} - q's'\{yy\} \qquad \text{(II.75)}$$

$$\boldsymbol{D(3)}$$

The multiplet formalism { } can help us only so far. It does not apply to the coupled evolution of terms like $\{zx\}$ or $\{yx\}$ since this coupled evolution is followed by some more pulses (cf. Rule #5 in Appendix L). We have to consider separately the CH, CH_2, and CH_3, using the corresponding subscripts 1,2,3. For completeness, we will also consider the case of the nonprotonated carbon (subscript zero).

$$D_o(3) = \lambda p'[y]$$

$$D_1(3) = \lambda p'[y1] + q'c'[yx] - q's'[yy]$$

$$D_2(3) = \lambda p'[y11] + q'c'([yx1] + [y1x]) - q's'([yy1] + [y1y])$$

$$D_3(3) = \lambda p'[y111] + q'c'([yx11] + [y1x1] + [y11x]) - q's'([yy11] + [y1y1] + [y11y]) \qquad \text{(II.76)}$$

The evolution from $t(3)$ to $t(4)$ leads to

$$D_o(3) \xrightarrow{\text{shift C}} \lambda p'(c[y]-s[x]) = D_o(4) \qquad \text{(no coupling)}$$

$$D_1(3) \xrightarrow{J\text{ coupl.}} -\lambda p'[xz]+q'(c'[yx]-s'[yy])$$

$$\xrightarrow{\text{shift H}} -\lambda p'[xz]+q'(c'^2[yx]+c's'[yy]-s'c'[yy]+s'^2[yx])$$

$$= -\lambda p'[xz]+q'[yx]$$

$$\xrightarrow{\text{shift C}} -\lambda p'(c[xz]+s[yz])+q'(c[yx]-s[xx]) = D_1(4)$$

In processing the evolution of $D_2(3)$ we have to treat separately the coupling of the carbon with the first and with the second proton.

$$D_2(3) \xrightarrow{\text{cpl AX1}} -\lambda p'[xz1]+q'c'([yx1]-[xzx])-q's'([yy1]-[xzy])$$

$$\xrightarrow{\text{cpl AX2}} -\lambda p'[yzz]+q'c'(-[xxz]-[xzx])-q's'([xyz]-[xzy])$$

$$\xrightarrow{\text{shift H}} -\lambda p'[yzz]+q'c'^2(-[xxz]-[xzx])+q'c's'(-[xyz]-[xzy])$$
$$-q's'c'(-[xyz]-[xzy])-q's'^2([xxz]+[xzx])$$

$$= -\lambda p'[yzz]-q'([xxz]+[xzx]) \xrightarrow{\text{shift C}} -\lambda p'(c[yzz]-s[xzz])$$
$$-q'c([xxz]+[xzx])-q's([yxz]+[yzx]) = D_2(4)$$

We were allowed to handle the evolution of both protons in one step only because none of the POs had x or y for both protons, i.e., only one proton was affected by the evolution in each PO.

Similar calculations will produce $D_3(4)$. Summarizing the results at $t(4)$ we have:

$$D_o(4) = \lambda p'(c[y]-s[x])$$

$$D_1(4) = \lambda p'(-c[xz]-s[yz])+q'c[yx]-q's[xx] \qquad \text{(II.77)}$$

$$D_2(4) = \lambda p'(-c[yzz]+s[xzz])-q'c([xxz]+[xzx])-q's([yxz]+[yzx])$$

$$D_3(4) = \lambda p'(c[xzzz]+s[yzzz])-q'c([yxzz]+[yzxz]+[yzzx])$$
$$+q's([xxzz]+[xzxz]+[xzzx])$$

The $\theta \pm y$ pulse is mathematically equivalent to a $\pm\theta\, y$ pulse. Therefore, we will treat it as a rotation about the y-axis with alternate signs of θ. We take also into account that

$$\cos(-\theta) = \cos\theta \quad ; \quad \sin(-\theta) = -\sin\theta$$
$$\cos(\pm\theta) = \cos\theta \quad ; \quad \sin(\pm\theta) = \pm\sin\theta$$

$$D_o(4) \xrightarrow{180xC} \lambda p'(-c[y] - s[x]) = D_o(5)$$

(the proton pulse has no effect).

$$D_1(4) \xrightarrow{180xC} \lambda p'(-c[xz] + s[yz]) - q'c[yx] - q's[xx]$$

$$\xrightarrow{\pm\theta yH} \lambda p'\cos\theta(-c[xz] + s[yz]) \pm q'\sin\theta(c[yz] + q's[xz] + \text{NOT}$$
$$= D_1(5)$$

In the last calculation we have retained only the observable terms ([x1],[y1])and the potentially observable terms ([xz],[yz]). We have relegated terms as [xx], [xy], [yx], [yy] to the NOT bunch (non-observable terms), as described in Appendix K.

In processing $D_2(4)$ we have to calculate the effect of the proton pulse, which is neither 90° nor 180°, separately on the two protons (see end of Section II.6).

$$D_2(4) \xrightarrow{180xC}$$

$$\longrightarrow \lambda p'(c[yzz] + s[xzz]) - q'c([xxz] + [xzx]) + q's([yxz] + [yzx])$$

$$\xrightarrow{\pm\theta yX1} \lambda p'\cos\theta(c[yzz] + s[xzz]) + q'\cos\theta(-c[xzx] + s[yzx])$$
$$\pm q'\sin\theta(c[xzz] - s[yzz]) + \text{NOT}$$

$$\xrightarrow{\pm\theta yX2} \lambda p'\cos^2\theta(c[yzz] + s[xzz]) \pm q'\cos\theta\sin\theta(c[xzz] - s[yzz])$$
$$\pm q'\sin\theta\cos\theta(c[xzz] - s[yzz]) + \text{NOT}$$

$$= \lambda p'\cos^2\theta(c[yzz] + s[xzz]) \pm 2q'\cos\theta\sin\theta(c[xzz] - s[yzz]) + \text{NOT}$$
$$= D_2(5)$$

Similar calculations have to be performed on $D_3(4)$ and the situation at $t(5)$ is

$$D_o(5) = \lambda p'(-c[y] - s[x])$$

$$D_1(5) = \lambda p' \cos\theta(-c[xz] + s[yz]) \pm q' \sin\theta(c[yz] + q' s[xz] + \mathrm{NOT}$$

$$\begin{aligned} D_2(5) &= \lambda p' \cos^2\theta(c[yzz] + s[xzz]) \\ &\pm 2q' \cos\theta \sin\theta(c[xzz] - s[yzz]) + \mathrm{NOT} \end{aligned} \qquad (II.78)$$

$$\begin{aligned} D_3(5) &= lp' \cos^3 q(c[xzzz] - s[yzzz]) \\ &\pm 3q' \cos^2 q \sin q(-c[yzzz] - s[xzzz]) + NOT \end{aligned}$$

Follows now the last $1/2J$ coupled evolution. We will retain only the observable terms, having x or y for carbon and 1 for all protons.

$$\begin{aligned} D_o(5) &\xrightarrow{\text{shift } C} \lambda p'(-c^2[y] + cs[x] - sc[x] - s^2[y]) = -\lambda p'[y] \\ &= -\lambda p'[y] = D_o(6) \end{aligned}$$

$$\begin{aligned} D_1(5) &\xrightarrow{\text{coupl.}} \lambda p' \cos\theta(-c[y1] - s[x1]) \pm q' \sin\theta(-c[x1] + s[y1] + \mathrm{NOT} \\ &\xrightarrow{\text{shift C}} \lambda p' \cos\theta(-c^2[y1] + cs[x1] - sc[x1] - s^2[y1]) \\ &\pm q' \sin\theta(-c^2[x1] - cs[y1] + sc[y1] - s^2[x1]) + \mathrm{NOT} \\ &= -\lambda p' \cos\theta[y1] - (\pm q' \sin\theta[x1]) + \mathrm{NOT} = D_1(6) \end{aligned}$$

After performing similar calculatons for $D_2(5)$ and $D_3(5)$, we can summarize the results at $t(6)$ as follows

$$\begin{aligned} D_o(6) &= -\lambda p'[y] \\ D_1(6) &= -(\lambda p' \cos\theta[y1] \pm q' \sin\theta[x1]) + \mathrm{NOT} \\ D_2(6) &= -(\lambda p' \cos^2\theta[y11] \pm 2q' \sin\theta \cos\theta[x11]) + \mathrm{NOT} \\ D_3(6) &= -(\lambda p' \cos^3\theta[y111] \pm 3q' \sin\theta \cos^2\theta[x111]) + \mathrm{NOT} \end{aligned} \qquad (II.79)$$

As a remarkable achievement of the DEPT sequence, we notice that no chemical shift (proton or carbon) is expressed in the density matrix at time $t(6)$, when the acquisition begins. We will not have any frequency dependent phase shift. This alone can justify the "distortionless" claim in the name of the sequence.

The term containing λ can be relatively small when a high repetition rate is used. We are interested in the second term, which does not contain the factor λ and is also polarization enhanced (has q' rather than p').

The first term can be edited out by phase cycling. The $\pm\sin\theta$ factor in the expression of the density matrix corresponds to the last proton pulse (see Figure II.8) being applied along the $+y$ and $-y$ axis, respectively. If we take a scan with the phase $+y$ and subtract it from the one with phase $-y$, the first term is cancelled and we are left with

$$
\begin{aligned}
D_1(6) &= q' \sin\theta[x1] + \mathrm{NOT} \\
D_2(6) &= 2q' \sin\theta\cos\theta[x11] + \mathrm{NOT} \\
D_3(6) &= 3q' \sin\theta\cos^2\theta[x111] + \mathrm{NOT}
\end{aligned}
\tag{II.80}
$$

The subtraction is performed by a 180° shift in the receiver phase. Theoretically, a two-step phase cycling is enough, in which the phase of the last proton pulse is

$-y$ $+y$

and the receiver phase is

0° 180°

Additional phase cycling is commonly used to cancel radio-frequency interferences and the effect of pulse imperfections.

The nonprotonated carbons do not appear in the phase-cycled spectrum. The discrimination between CH, CH_2, and CH_3 is done by running the sequence three times, with different values of the flip angle θ, namely 90°, 45° and 135°.

When $\theta = 90°$, then $\cos\theta = 0$, $\sin\theta = 1$ and (II.80) becomes

$$D_1(6) = q'[x1]) + \mathrm{NOT}$$

$$D_2(6) = \mathrm{NOT}$$

$$D_3(6) = \mathrm{NOT}$$

This is a ^{13}C spectrum in which only the CH lines appear. When θ is slightly larger or smaller than 90°, a CH_2 will appear as a small singlet, negative or positive, respectively. That is why DEPT is a good method for calibrating the 90° proton pulse in a spectrometer configured for observing carbon.

When θ = 45°, cosθ = $1/\sqrt{2}$, sinθ = $1/\sqrt{2}$ and (II.80) becomes

$$D_1(6) = (1/\sqrt{2})q'[x1]) + \text{NOT}$$

$$D_2(6) = q'[x11] + \text{NOT}$$

$$D_3(6) = (3/2\sqrt{2})\,q'[x111]) + \text{NOT}$$

All CH, CH_2, and CH_3 appear as positive singlets.

When θ = 135°, then cosθ = $-1/\sqrt{2}$, sinθ = $1/\sqrt{2}$ and (II.80) becomes

$$D_1(6) = (1/\sqrt{2})q'[x1]) + \text{NOT}$$

$$D_2(6) = -\, q'[x11] + \text{NOT}$$

$$D_3(6) = (3/2\sqrt{2})\,q'[x111]) + \text{NOT}$$

This is the same situation as for 45°, but CH_2 peaks appear negative.

We have discussed all the features of DEPT using the expression of the density matrix at time $t(6)$, when the acquisition begins. The ^{13}C magnetization, which is oriented along x at $t(6)$, will precess during the acquisition with the Larmor frequency corresponding to the respective line and will appear as a singlet.

In principle only the 90° and the 135° runs are sufficient for an unambigous identification of the CH, CH_2, and CH_3 peaks. The 45° run is necessary when one wants to do spectral editing. For the nonprotonated carbons, a normal run is needed with a long recycle time.

Modern instruments offer the possibility to edit DEPT spectra in order to select only the peaks of one group (e.g., CH) while eliminating the peaks of the other two groups (CH_2 and CH_3). This is theoretically based on linear combinations of the spectra obtained with different values of θ (45°, 90°, 135°).

Table II.2 contains the relative amplitudes of CH, CH_2, and CH_3 peaks for each of the three angles.

Table II.2. Relative peak amplitudes in the raw DEPT spectra.

Spectrum	θ	CH	CH_2	CH_3
A	45°	$1/\sqrt{2}$ (0.707)	1	$3/(2\sqrt{2})$ (1.06)
B	90°	1	0	0
C	135°	$1/\sqrt{2}$ (0.707)	–1	$3/(2\sqrt{2})$ (1.06)

Table II.3 shows the operations necessary to obtain spectrs of only one of the three groups or all of them together (note that the spectrum taken at 45° also shows the peaks of all groups, but not their true amplitudes).

Table II.3. Linear combinations of the raw DEPT spectra, necessary in order to obtain only one of the three groups or all of them together.

Combination	CH	CH_2	CH_3
B	1	0	0
(A – C)/2	0	1	0
(A+C - 1.41B)/2.12	0	0	1
Sum of the above	1	1	1

Some instrument softwares make it possible to alter the theoretical coefficients in order to compensate for hardware imperfections.

14. PO TREATMENT OF APT (ATTACHED PROTON TEST)

APT is probably the first sequence designed to identify the number of hydrogen atoms attached to each carbon in a decoupled ^{13}C spectrum. The sequence is relatively simple and does not involve proton pulses. The broadband proton decoupler is kept permanently on, with the exception of one specific 1/J delay.

The sequence can be correctly analyzed within the frame of the vector representation. Here we treat it with the PO formalism. The sequence is shown in Figure II.9.

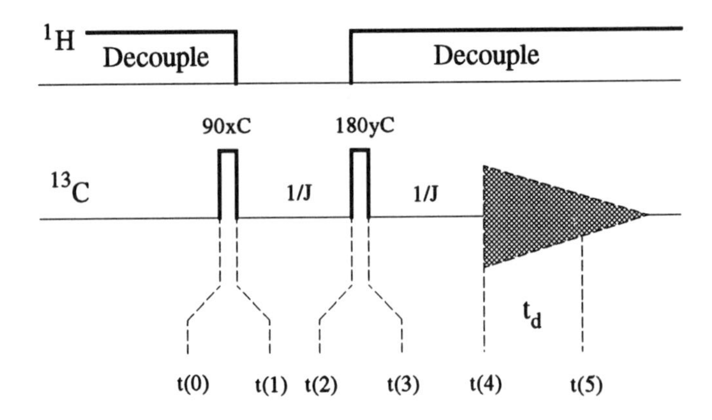

Figure II.9. The APT sequence: $90xC - 1/J - 180yC \; - 1/J$ (decouple) $-$ AT(decouple)

We will treat separately the C, CH, CH_2, and CH_3 systems, denoting the respective density matrices with subscripts 0 to 3.

$$D_o(0) = -p'[z]$$

$$D_1(0) = -p'[z1] - q'[1z] \quad \text{(II.81)}$$

$$D_2(0) = -p'[z11] - q'([1z1] + [11z])$$

$$D_3(0) = -p'[z111] - q'([1z11] + [11z1] + [111z])$$

Since no coherent proton pulse is applied throughout the sequence, we can anticipate that the proton POs will never turn into ^{13}C observables. We can treat them as NOT (nonobservable terms). We start therefore with

$$\begin{aligned} D_o(0) &= -p'[z] \\ D_1(0) &= -p'[z1] + \text{NOT} \\ D_2(0) &= -p'[z11] + \text{NOT} \\ D_3(0) &= -p'[z111] + \text{NOT} \end{aligned} \qquad \text{(II.82)}$$

Anyhow, the proton components do not have a predictable behavior during the delay in which the decoupler is turned on (see Figure II.9).

$$\begin{aligned} D_o(0) &\xrightarrow{90xC} p'[y] = D_o(1) \\ D_1(0) &\xrightarrow{90xC} p'[y1] + \text{NOT} = D_1(1) \\ D_2(0) &\xrightarrow{90xC} p'[y11] + \text{NOT} = D_2(1) \\ D_3(0) &\xrightarrow{90xC} p'[y111] + \text{NOT} = D_3(1) \end{aligned} \qquad \text{(II.83)}$$

For the evolution $1/J$ we use the notations

$$c = \cos \Omega_C / J \quad ; \quad C = \cos \pi J / J = \cos \pi = -1$$
$$s = \sin \Omega_C / J \quad ; \quad S = \sin \pi J / J = \sin \pi = 0$$

$$D_o(1) \xrightarrow{\text{shift}\,C} p'(c[y] - s[x]) = D_o(2)$$

$$D_1(1) \xrightarrow{\text{coupl.}} -p'[y1] + \text{NOT}$$
$$\xrightarrow{\text{shift}\,C} -p'(c[y1] - s[x1]) + \text{NOT} = D_1(2)$$

$$D_2(1) \xrightarrow{\text{coupl.}\,AX1} -p'[y11] + \text{NOT} \xrightarrow{\text{coupl.}\,AX2} +p'[y11] + \text{NOT}$$
$$\xrightarrow{\text{shift}\,C} p'(c[y11] - s[x11]) + \text{NOT} = D_2(2)$$

$$D_3(1) \xrightarrow{\text{coupl.}\,AX1} -p'[y111] + \text{NOT} \xrightarrow{\text{coupl.}\,AX2} +p'[y111] + \text{NOT}$$
$$\xrightarrow{\text{coupl.}\,AX3} -p'[y111] + \text{NOT}$$
$$\xrightarrow{\text{shift}\,C} -p'(c[y111] - s[x111]) + \text{NOT} = D_3(2)$$

In (II.84) we summarize the results at $t(2)$. It is not necessary to carry the NOT any farther.

$$\begin{aligned} D_o(2) &= +p'\,(c[y]-s[x]) \\ D_1(2) &= -p'\,(c[y1]-s[x1]) \\ D_2(2) &= +p'\,(c[y11]-s[x11]) \\ D_3(2) &= -p'\,(c[y111]-s[x111]) \end{aligned} \qquad \text{(II.84)}$$

We observe that the density matrices for CH and CH_3 have a minus sign. It comes from the fact that, during the evolution $1/J$, the coupling to each proton reverses the sign (C = -1). This is exactly what the APT is meant to do. The following pulse and delay are only intended to refocus the shift evolution, in order to avoid frequency dependent phaseshifts.

$$\begin{aligned} D_o(2) &\xrightarrow{180yC} +p'\,(c[y]+s[x]) = D_o(3) \\ D_1(2) &\xrightarrow{180yC} -p'\,(c[y1]+s[x1]) = D_1(3) \\ D_2(2) &\xrightarrow{180yC} +p'\,(c[y11]+s[x11]) = D_2(3) \\ D_3(2) &\xrightarrow{180yC} -p'\,(c[y111]+s[x111]) = D_3(3) \end{aligned}$$

Now we have the $1/J$ coupled evolution

$$D_o(3) \xrightarrow{\text{shift}\,C} +p'\,(c^2[y]-cs[x]+sc[x]+s^2[y]) = +p'\,[y] = D_o(4)$$

It works the same way for the other three cases and at t(4) we have

$$\begin{aligned} D_o(4) &= +p'\,[y] \\ D_1(4) &= -p'\,[y1] \\ D_2(4) &= +p'\,[y11] \\ D_3(4) &= -p'\,[y111] \end{aligned} \qquad \text{(II.84)}$$

At $t(4)$ the detection starts. All peaks will be singlets (decoupled acquisition) but CH and CH_3 will be negative. Admittedly, the PO treatment is less instructive about how this has been achieved than the vector representation.

Even with the sequences that cannot be handled with the vector representation alone, it may be instructive to follow the evolution of the magnetization components as indicated by the density matrix expressions. We have done this in Section 3.11 of Part I.

APPENDIX A: MATH REMINDER

Complex numbers

A complex number

$$z = x + iy \tag{A1}$$

can be graphically represented as in Figure A.1, where

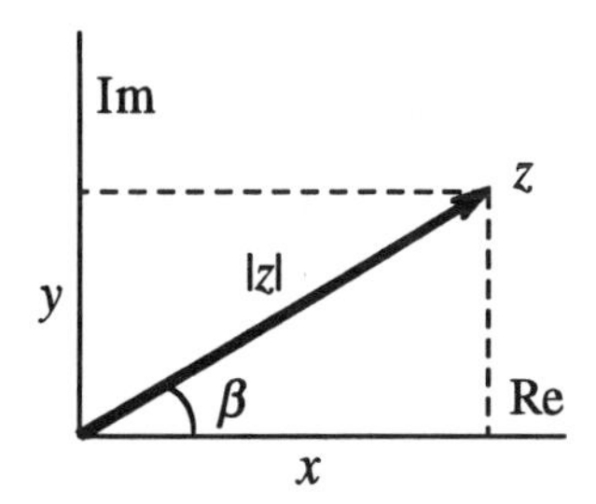

Im is the imaginary axis
Re is the real axis
x is the real part of z
y is the coefficient of the imaginary part of z
$|z|$ is the modulus (absolute value) of z
β is the argument of z

Figure A.1. Graphic representation of a complex number $z = x + iy$. $i = \sqrt{-1}$ is the imaginary unit.

The number z is fully determined when either x and y or $|z|$ and β are known. The relations between these two pairs of variables are (see lower triangle):

$$x = |z| \cos \beta \tag{A2}$$

$$y = |z| \sin \beta \tag{A3}$$

Thus, according to (A1),

$$z = |z| \cos \beta + i|z| \sin \beta = |z|(\cos \beta + i \sin \beta) \tag{A4}$$

Using Euler's formula [see (A11)-(A16)], one obtains:

$$z = |z| \exp(i\beta) \tag{A5}$$

The complex number $z^* = x - iy = |z| \exp(-i\beta)$ is called the *complex* conjugate of z.

Elementary rotation operator

Consider a complex number r which has a modulus $|r| = 1$ and the argument α:

$$r = \exp(i\alpha) = \cos\alpha + i\sin\alpha \tag{A6}$$

Multiplying a complex number such as

$$z = |z|\exp(i\beta)$$

by r leaves the modulus of z unchanged and increases the argument by α:

$$zr = |z|e^{i\beta}e^{ia} = |z|e^{i(\beta+\alpha)} \tag{A7}$$

Equation (A7) describes the rotation of the vector Oz by an angle α (see Figure A.2). We can call r, the *elementary rotation operator.*

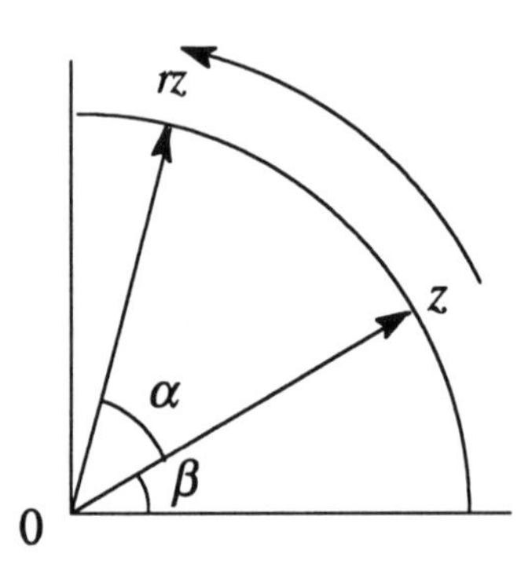

Figure A.2. Effect of the rotation operator $r = \exp(i\alpha)$ on the complex number $z = |z|\exp(i\beta)$.

Note: Although more complicated, the rotation operators in the density matrix treatment of multipulse NMR are of the same form as our elementary operator [cf.(B45)].

Example 1. A $-90°$ (clockwise) rotation (see Figure A.3)

Let $$\beta = 90° \quad ; \quad \alpha = -90°$$

Then, from (A4 and A6),

$$z = |z|(\cos 90° + i \sin 90°) = i|z| \tag{A8}$$

$$r = \cos(-90°) + i \sin(-90°) = -i \tag{A9}$$

The product

$$zr = i|z|(-i) = |z| \tag{A10}$$

is a real number (its argument is zero).

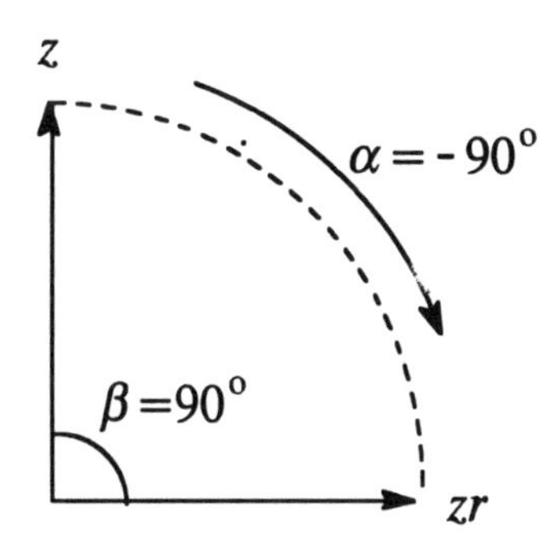

Figure A.3. Effect of the rotation operator . $-i$ $\left(\alpha = -90°\right)$

Equation (A10) tells us that the particular operator $-i$ effects a 90° CW (clockwise) rotation on the vector z. The operator $+i$ would rotate z by 90° CCW. Two consecutive multplications by i result in a 180° rotation. In other words, an i^2 operator (-1) orients the vector in opposite direction.

Example 2. Powers of i (the "star of i")

Since i represents a 90° CCW rotation, successive powers of i are obtained by successive 90° CCW rotations (see Figure A.4).

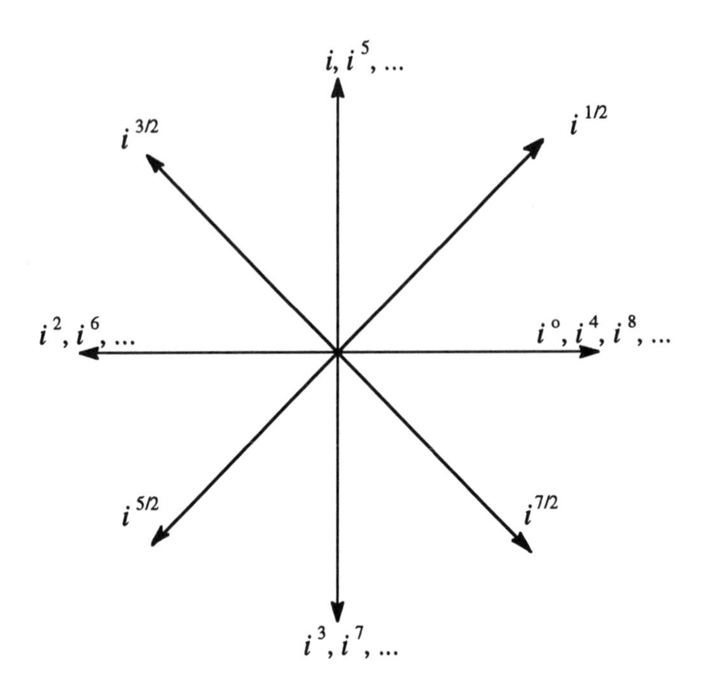

Figure A.4. The star of i.

Series expansion of e^x, $\sin x$, $\cos x$, and e^{ix}

The exponential function describes most everything happening in nature (including the evolution of your investment accounts). The e^x series is

$$e^x = 1 + x + x^2/2! + x^3/3! + \ldots \tag{A11}$$

The sine and cosine series are:

$$\sin x = x - x^3/3! + x^5/5! - x^7/7! + \ldots \tag{A12}$$

$$\cos x = 1 - x^2/2! + x^4/4! - x^6/6! + \ldots \tag{A13}$$

The e^{ix} series is:

$$e^{ix} = 1 + ix - x^2/2! - ix^3/3! + x^4/4! + ix^5/5! \ldots \quad \text{(A14)}$$

Separation of the real and imaginary terms gives

$$e^{ix} = 1 - x^2/2! + x^4/4! - x^6/6! + \ldots \quad \text{(A15)}$$
$$+ i(x - x^3/3! + x^5/5! - \ldots)$$

Recognizing the sine and cosine series one obtains the Euler formula

$$e^{ix} = \cos x + i \sin x \quad \text{(A16)}$$

Matrix algebra

Matrices are arrays of elements disposed in rows and columns; they obey specific algebraic rules for addition, multiplication and inversion. The following formats of matrices are used in quantum mechanics:

$$A = \begin{bmatrix} a_{11} & a_{12} & . & . & a_{1n} \\ a_{21} & a_{22} & . & . & a_{2n} \\ . & . & . & . & . \\ . & . & . & . & . \\ a_{n1} & a_{n2} & . & . & a_{nn} \end{bmatrix}$$

Square matrix

$$A = \begin{bmatrix} a_{11} & a_{12} & . & . & a_{1n} \end{bmatrix} \quad ; \quad A = \begin{bmatrix} a_{11} \\ a_{21} \\ . \\ . \\ a_{n1} \end{bmatrix}$$

Row matrix *Column matrix*

Row and column matrices are also called row vectors or column vectors. Note that the first subscript of each element indicates the row number and the second, the column number.

Matrix addition

Two matrices are added element by element as follows:

$$\begin{bmatrix} a_{11} & a_{12} & a_{13} \\ a_{21} & a_{22} & a_{23} \\ a_{31} & a_{32} & a_{33} \end{bmatrix} + \begin{bmatrix} b_{11} & b_{12} & b_{13} \\ b_{21} & b_{22} & b_{23} \\ b_{31} & b_{32} & b_{33} \end{bmatrix} = \begin{bmatrix} a_{11}+b_{11} & a_{12}+b_{12} & a_{13}+b_{13} \\ a_{21}+b_{21} & a_{22}+b_{22} & a_{23}+b_{23} \\ a_{31}+b_{31} & a_{32}+b_{32} & a_{33}+b_{33} \end{bmatrix}$$

Only matrices with the same number of rows and columns can be added.

Matrix multiplication

In the following are shown three typical matrix multiplications.

a) Square matrix times column matrix:

$$\begin{bmatrix} a_{11} & a_{12} & a_{13} \\ a_{21} & a_{22} & a_{23} \\ a_{31} & a_{32} & a_{33} \end{bmatrix} \times \begin{bmatrix} b_{11} \\ b_{21} \\ b_{31} \end{bmatrix} = \begin{bmatrix} c_{11} \\ c_{21} \\ c_{31} \end{bmatrix}$$

$$\begin{aligned} c_{11} &= a_{11}b_{11} + a_{12}b_{21} + a_{13}b_{31} \\ c_{21} &= a_{21}b_{11} + a_{22}b_{21} + a_{23}b_{31} \\ c_{31} &= a_{31}b_{11} + a_{32}b_{21} + a_{33}b_{31} \end{aligned} \qquad \text{i.e.,} \quad c_{j1} = \sum_{k=1}^{3} a_{jk}b_{k1} \qquad (j = 1, 2, 3)$$

Example:

$$\begin{bmatrix} 1 & 2 & 3 \\ 4 & 5 & 6 \\ 7 & 8 & 9 \end{bmatrix} \times \begin{bmatrix} 1 \\ 2 \\ 3 \end{bmatrix} = \begin{bmatrix} 14 \\ 32 \\ 50 \end{bmatrix}$$

Each element of the first line of the square matrix has been multiplied with the corresponding element of the column matrix to yield the first element of the product:

$$1 \times 1 + 2 \times 2 + 3 \times 3 = 14$$

b) Row matrix times square matrix:

$$\begin{bmatrix} a_{11} & a_{12} & a_{13} \end{bmatrix} \times \begin{bmatrix} b_{11} & b_{12} & b_{13} \\ b_{21} & b_{22} & b_{23} \\ b_{31} & b_{32} & b_{33} \end{bmatrix} = \begin{bmatrix} c_{11} & c_{12} & c_{13} \end{bmatrix}$$

$$\begin{aligned} c_{11} &= a_{11}b_{11} + a_{12}b_{21} + a_{13}b_{31} \\ c_{12} &= a_{11}b_{12} + a_{12}b_{22} + a_{13}b_{32} \\ c_{13} &= a_{11}b_{13} + a_{12}b_{23} + a_{13}b_{33} \end{aligned} \quad \text{i.e.,} \quad c_{1j} = \sum_{k=1}^{3} a_{1k}b_{kj} \qquad (j = 1,2,3)$$

c) Square matrix times square matrix:

$$\begin{bmatrix} a_{11} & a_{12} & a_{13} \\ a_{21} & a_{22} & a_{23} \\ a_{31} & a_{32} & a_{33} \end{bmatrix} \times \begin{bmatrix} b_{11} & b_{12} & b_{13} \\ b_{21} & b_{22} & b_{23} \\ b_{31} & b_{32} & b_{33} \end{bmatrix} = \begin{bmatrix} c_{11} & c_{12} & c_{13} \\ c_{21} & c_{22} & c_{23} \\ c_{31} & c_{32} & c_{33} \end{bmatrix}$$

$$c_{jm} = \sum_{k=1}^{3} a_{jk}b_{km}$$

For instance, row 2 of the left hand matrix and column 3 of the right hand matrix are involved in the obtaining of the element c_{23} of the product. Example:

$$\frac{1}{\sqrt{2}}\begin{bmatrix} 0 & 1 & 0 \\ 1 & 0 & 1 \\ 0 & 1 & 0 \end{bmatrix} \times \frac{1}{\sqrt{2}}\begin{bmatrix} 0 & -i & 0 \\ i & 0 & -i \\ 0 & i & 0 \end{bmatrix} =$$

$$\frac{1}{2}\begin{bmatrix} i & 0 & -i \\ 0 & 0 & 0 \\ i & 0 & -i \end{bmatrix} = \begin{bmatrix} i/2 & 0 & -i/2 \\ 0 & 0 & 0 \\ i/2 & 0 & -i/2 \end{bmatrix}$$

The product inherits the number of rows from the first (left) matrix and the number of columns from the second (right) matrix. The number of columns of the left matrix must match the number of rows of the right matrix.

In general the matrix multiplication is not commutative:

$$AB \neq BA \tag{A17}$$

It is associative:

$$A(BC) = (AB)C = ABC \tag{A18}$$

and distributive:

$$A(B + C) = AB + AC \tag{A19}$$

The *unit matrix* is shown below

$$[\mathbf{1}] = \begin{bmatrix} 1 & 0 & 0 & 0 \\ 0 & 1 & 0 & 0 \\ 0 & 0 & 1 & 0 \\ 0 & 0 & 0 & 1 \end{bmatrix} \tag{A20}$$

It must be square and it can be any size. Any matrix remains unchanged when multiplied with the unit matrix:

$$[\mathbf{1}] \times A = A \times [\mathbf{1}] = A \tag{A21}$$

Matrix inversion

The inverse A^{-1} of a square matrix A is defined by the relation:

$$A \times A^{-1} = A^{-1} \times A = [\mathbf{1}] \tag{A22}$$

To find A^{-1}:

1) Replace each element of A by its signed minor determinant. The minor determinant of the matrix element a_{jk} is built with the elements of the original matrix after striking out row j and column k. To have the *signed* minor determinant, one has to multiply it by – 1 whenever the sum j+k is odd.

2) Interchange the rows and the columns (this operation is called matrix transposition)

3) Divide all elements of the transposed matrix by the determinant of the original matrix A.

Example:
Find the inverse of matrix A.

$$A = \frac{1}{2}\begin{bmatrix} 1 & \sqrt{2} & 1 \\ -\sqrt{2} & 0 & \sqrt{2} \\ 1 & -\sqrt{2} & 1 \end{bmatrix} = \begin{bmatrix} 1/2 & \sqrt{2}/2 & 1/2 \\ -\sqrt{2}/2 & 0 & \sqrt{2}/2 \\ 1/2 & -\sqrt{2}/2 & 1/2 \end{bmatrix}$$

1) We replace a_{11} by

$$\begin{vmatrix} a_{22} & a_{23} \\ a_{32} & a_{33} \end{vmatrix} = \begin{vmatrix} 0 & \sqrt{2}/2 \\ -\sqrt{2}/2 & 1/2 \end{vmatrix} = 0 \times (1/2) + (\sqrt{2}/2) \times (\sqrt{2}/2) = 1/2$$

This is the minor determinant obtained by striking out row 1 and column 1 of the original matrix.

We replace a_{12} by

$$-\begin{vmatrix} a_{21} & a_{23} \\ a_{31} & a_{33} \end{vmatrix} = -\begin{vmatrix} -\sqrt{2}/2 & \sqrt{2}/2 \\ 1/2 & 1/2 \end{vmatrix}$$

$$= -[(-\sqrt{2}/2) \times (1/2) - (\sqrt{2}/2) \times (1/2)] = \sqrt{2}/2$$

and so on, obtaining:

$$\begin{bmatrix} 1/2 & \sqrt{2}/2 & 1/2 \\ -\sqrt{2}/2 & 0 & \sqrt{2}/2 \\ 1/2 & -\sqrt{2}/2 & 1/2 \end{bmatrix}$$

2) The transposition yields:

$$\begin{bmatrix} 1/2 & -\sqrt{2}/2 & 1/2 \\ \sqrt{2}/2 & 0 & -\sqrt{2}/2 \\ 1/2 & \sqrt{2}/2 & 1/2 \end{bmatrix}$$

3) Calculate the determinant of A:

$$\det(A) = 0 + (1/4) + (1/4) - 0 + (1/4) + (1/4) = 1$$

Divided by 1, the transposed matrix remains unchanged:

$$A^{-1} = \begin{bmatrix} 1/2 & -\sqrt{2}/2 & 1/2 \\ \sqrt{2}/2 & 0 & -\sqrt{2}/2 \\ 1/2 & \sqrt{2}/2 & 1/2 \end{bmatrix} = \frac{1}{2}\begin{bmatrix} 1 & -\sqrt{2} & 1 \\ \sqrt{2} & 0 & -\sqrt{2} \\ 1 & \sqrt{2} & 1 \end{bmatrix}$$

Check: $$A^{-1} \times A = [\mathbf{1}]$$

$$\frac{1}{2}\begin{bmatrix} 1 & -\sqrt{2} & 1 \\ \sqrt{2} & 0 & -\sqrt{2} \\ 1 & \sqrt{2} & 1 \end{bmatrix} \times \frac{1}{2}\begin{bmatrix} 1 & \sqrt{2} & 1 \\ -\sqrt{2} & 0 & \sqrt{2} \\ 1 & -\sqrt{2} & 1 \end{bmatrix}$$

$$= \frac{1}{4}\begin{bmatrix} 4 & 0 & 0 \\ 0 & 4 & 0 \\ 0 & 0 & 4 \end{bmatrix} = \begin{bmatrix} 1 & 0 & 0 \\ 0 & 1 & 0 \\ 0 & 0 & 1 \end{bmatrix} = [\mathbf{1}]$$

Note: You will be pleased to learn that:

a) There is a shortcut for the inversion of rotation operator matrices because they are of a special kind.

b) In our calculations we will need only inversions of rotation operators:

$$(R \rightarrow R^{-1})$$

Here is the shortcut:

1) Transpose R (vide supra)

2) Replace each element with its complex conjugate.

Example

$$R = \frac{1}{2}\begin{bmatrix} 1 & \sqrt{2} & 1 \\ -\sqrt{2} & 0 & \sqrt{2} \\ 1 & -\sqrt{2} & 1 \end{bmatrix}$$

1) Transpose:

$$\frac{1}{2}\begin{bmatrix} 1 & -\sqrt{2} & 1 \\ \sqrt{2} & 0 & -\sqrt{2} \\ 1 & \sqrt{2} & 1 \end{bmatrix}$$

2) Conjugate: Because they are all real, the matrix elements remain the same:

$$R^{-1} = \frac{1}{2}\begin{bmatrix} 1 & -\sqrt{2} & 1 \\ \sqrt{2} & 0 & -\sqrt{2} \\ 1 & \sqrt{2} & 1 \end{bmatrix}$$

Note: This is the same matrix as in the previous example; its being a rotation operator matrix, allowed us to use the shortcut procedure.

Another example:

$$R = \frac{1}{\sqrt{2}}\begin{bmatrix} 1 & 0 & i & 0 \\ 0 & 1 & 0 & i \\ i & 0 & 1 & 0 \\ 0 & i & 0 & 1 \end{bmatrix}$$

(1) Transpose: you obtain the same matrix

(2) Conjugate:

$$R^{-1} = \frac{1}{\sqrt{2}} \begin{bmatrix} 1 & 0 & -i & 0 \\ 0 & 1 & 0 & -i \\ -i & 0 & 1 & 0 \\ 0 & -i & 0 & 1 \end{bmatrix}$$

Note: The matrix resulting from the transposition followed by complex conjugation of a given matrix A is called the adjoint matrix A^{adj}. For all rotation operators,

$$R^{-1} = R^{adj} \tag{A23}$$

In other words, our short cut for inversion is equivalent with finding the adjoint of R [see (A23)].

When $$A = A^{adj}$$

we say that the matrix A is *self adjoint* or *Hermitian*. In a Hermitian matrix, every element below the main diagonal is the complex conjugate of its symmetrical element above the diagonal

$$d_{nm} = d^*_{mn} \tag{A24}$$

while the diagonal elements are all real. The angular momentum, the Hamiltonian and density matrix are all Hermitian (the rotation operators never are).

The matrix algebra operations encountered in Part 1 of this book are mostly multiplications of square matrices (density matrix and product operators). Inversion is only used to find reciprocals of rotation operators, by transposition and conjugation. Frequently used is the multiplication or division of a matrix by a constant, performed by multiplying or dividing every element of the matrix. The derivative of a matrix (e.g., with respect to time) is obtained by taking the derivative of each matrix element .

Trigonometric relations

Sum of squared sine and cosine

$$\sin^2\alpha + \cos^2\alpha = 1 \tag{A25}$$

Negative angles

$$\begin{aligned} \sin(-\alpha) &= -\sin\alpha \\ \cos(-\alpha) &= \cos\alpha \end{aligned} \tag{A26}$$

Expressing sine and cosine in terms of exponentials

$$\cos\alpha = \frac{e^{i\alpha} + e^{-i\alpha}}{2} \tag{A27}$$

$$\sin\alpha = \frac{e^{i\alpha} - e^{-i\alpha}}{2i} \tag{A28}$$

Demo: Use Euler's formula (A16) $e^{\pm i\alpha} = \cos\alpha \pm i\sin\alpha$

Sum and difference of two angles

$$\sin(\alpha \pm \beta) = \sin\alpha\cos\beta \pm \cos\alpha\sin\beta \tag{A29}$$

$$\cos(\alpha \pm \beta) = \cos\alpha\cos\beta \mp \sin\alpha\sin\beta \tag{A30}$$

Demo:

$$\begin{aligned} z_1 &= e^{i\alpha} = \cos\alpha + i\sin\alpha \\ z_2 &= e^{\pm i\beta} = \cos\beta \pm i\sin\beta \end{aligned}$$

$$\begin{aligned} z_1 z_1 = e^{i(\alpha \pm i\beta)} &= (\cos\alpha + i\sin\alpha)(\cos\beta \pm i\sin\beta) \\ &= (\cos\alpha\cos\beta \mp \sin\alpha\sin\beta) + i(\sin\alpha\cos\beta \pm \cos\alpha\sin\beta) \\ &= \cos(\alpha \pm \beta) + i\sin(\alpha \pm \beta) \end{aligned}$$

Angle 2α

$$\sin 2\alpha = 2\sin\alpha\cos\alpha \qquad \text{(A31)}$$

$$\cos 2\alpha = \cos^2\alpha - \sin^2\alpha = 2\cos^2\alpha - 1 \qquad \text{(A32)}$$

Demo: Make $\beta = \alpha$ in (A29),(A30). Use (A25) for the last form.

Angle 3α

$$\sin 3\alpha = 3\sin\alpha - 4\sin^3\alpha \qquad \text{(A33)}$$

$$\cos 3\alpha = 4\cos^3\alpha - 3\cos\alpha \qquad \text{(A34)}$$

Demo: Make β=2α in (A29),(A30), then use (A31),(A32) and eventually (A25).

Relations used for AX_2 systems

$$\cos^2\alpha = \frac{1+\cos 2\alpha}{2} = \frac{e^{2i\alpha} + 2 + e^{-2i\alpha}}{4} \qquad \text{(A35)}$$

$$\cos\alpha\sin\alpha = \frac{\sin 2\alpha}{2} = \frac{e^{2i\alpha} - e^{-2i\alpha}}{4i} \qquad \text{(A36)}$$

Demo: (A35) is a corollary of (A32) and (A36) is a corollary of (A31). Use (A27), (A28) to obtain the exponential form. Note the 1-2-1 triplet structure in (A 35).

Relations used for AX_3 systems

$$\cos^3\alpha = \frac{\cos 3\alpha + 3\cos\alpha}{4} = \frac{e^{3i\alpha} + 3e^{i\alpha} + 3e^{-i\alpha} + e^{-3i\alpha}}{8} \qquad \text{(A37)}$$

$$\cos^2\alpha\sin\alpha = \frac{\sin 3\alpha + 3\sin\alpha}{4} = \frac{e^{3i\alpha} + e^{i\alpha} - e^{-i\alpha} - e^{-3i\alpha}}{8i} \qquad \text{(A38)}$$

Demo: (A37) is a corollary of (A34). Relation (A38) can be obtained by rewriting (A33) as

$\sin 3\alpha = 3\sin\alpha - 4\sin\alpha(1-\cos^2\alpha) = 4\sin\alpha\cos^2\alpha - \sin\alpha$

Note the 1-3-3-1 quartet structure in (A37).

Relations used for AX systems

$$\cos\alpha\cos\beta = \frac{\cos(\alpha-\beta)+\cos(\alpha+\beta)}{2} \tag{A39}$$

$$\sin\alpha\sin\beta = \frac{\cos(\alpha-\beta)-\cos(\alpha+\beta)}{2} \tag{A40}$$

$$\sin\alpha\sin\beta = \frac{\cos(\alpha-\beta)-\cos(\alpha+\beta)}{2} \tag{A41}$$

$$\sin\alpha\cos\beta = \frac{\sin(\alpha-\beta)+\sin(\alpha+\beta)}{2} \qquad \text{(A42}$$

Demo: Introduce (A29) or (A30) in the second member of the equalities above.

APPENDIX B: DENSITY MATRIX FORMALISM

Wave functions and density matrix

The density matrix is a tool used to describe the state of a spin ensemble as well as its evolution in time. It allows the passage from the probabilistic treatment of a *system* of a few spins to the statistical treatment of a large *ensemble* of such systems.

Since we are interested in the magnetization we want to express this observable in terms of the wave function φ of the system. Let us concentrate on one of the nuclei in the system (e.g., nucleus A). The x component of the magnetic moment of nucleus A has the expectation value:

$$\langle \mu_{xA} \rangle = \langle \varphi | \mu_{xA} | \varphi \rangle = \gamma_A \hbar \langle \varphi | I_{xA} | \varphi \rangle \qquad \text{(B1)}$$

where I_{xA} is the operator of the x-component of the angular momentum of nucleus A in the given system. For instance, in an AX system the I_{xA} matrix is

$$\frac{1}{2}\begin{bmatrix} 0 & 1 & 0 & 0 \\ 1 & 0 & 0 & 0 \\ 0 & 0 & 0 & 1 \\ 0 & 0 & 1 & 0 \end{bmatrix} \qquad \text{[see (C12)]}$$

In order to calculate the macroscopic magnetization, we have to take the average (denoted by a bar) over the whole ensemble:

$$M_{xA} = N_0 \overline{\langle \mu_{xA} \rangle} = N_0 \gamma_A \hbar \overline{\langle \varphi | I_{xA} | \varphi \rangle} \qquad \text{(B2)}$$

where N_0 is the number of systems per unit volume, equal to the number of A spins per unit volume. Similar equations can be written for every component and for every nucleus in the system.

In the Schrödinger representation I_{xA} is a time independent operator, therefore the time dependence of M_{xA} is contained in the wave function φ of each system. This, in turn, may be expressed as a linear combination of the eigenstates $|n\rangle$ of the system:

$$\varphi = \sum_{n=1}^{N} c_n |n\rangle$$

N=number of quantum states of the system

Here again we observe that the eigenfunctions $|n\rangle$ are time independent (solutions of the time independent Schrödinger equation), so the time dependence is contained only in the coefficients c_n. In order to introduce these coefficients in the expression (B2), we put

$$\langle\varphi|I_{xA}|\varphi\rangle$$

in matrix form. The "ket" $|\varphi\rangle$ is a column matrix:

$$|\varphi\rangle = \begin{bmatrix} c_1 \\ c_2 \\ \cdot \\ \cdot \\ c_n \end{bmatrix}$$

The "bra" $\langle\varphi|$ is a row matrix

$$\langle\varphi| = \begin{bmatrix} c_1^* & c_2^* & \cdot & \cdot & c_n^* \end{bmatrix}$$

The angular momentum operator I_{xA} is an N by N square matrix.

$$\begin{bmatrix} I_{11} & I_{12} & \cdot & \cdot & I_{1N} \\ I_{21} & I_{22} & \cdot & \cdot & I_{2N} \\ \cdot & \cdot & \cdot & \cdot & \cdot \\ \cdot & \cdot & \cdot & \cdot & \cdot \\ I_{N1} & I_{N2} & \cdot & \cdot & I_{NN} \end{bmatrix}$$

The subscript "xA" has been omited to simplify the writing.

Using the expressions for $\langle \varphi |$, $|I_{xA}|$, and $| \varphi \rangle$ on the previous page, we obtain:

$$\langle \varphi | I_{xA} | \varphi \rangle = \begin{bmatrix} c_1^* & c_2^* & . & . & c_N^* \end{bmatrix}$$

$$\times \begin{bmatrix} I_{11} & I_{12} & . & . & I_{1N} \\ I_{21} & I_{22} & . & . & I_{2N} \\ . & . & . & . & . \\ . & . & . & . & . \\ I_{N1} & I_{N2} & . & . & I_{NN} \end{bmatrix} \times \begin{bmatrix} c_1 \\ c_2 \\ . \\ . \\ c_N \end{bmatrix}$$

$$= \begin{bmatrix} c_1^* & c_2^* & . & . & c_n^* \end{bmatrix} \times \begin{bmatrix} \sum_m I_{1m} c_m \\ \sum_m I_{2m} c_m \\ . \\ . \\ \sum_m I_{Nm} c_m \end{bmatrix} = \sum_n \sum_m c_n^* I_{nm} c_m = \sum_n \sum_m I_{nm} c_m c_n^* \tag{B4}$$

We have obtained a compact expression of

$$\langle \varphi | I_{xA} | \varphi \rangle$$

In order to introduce it in the expression (B2) of the magnetization we have to take its average over the whole ensemble of systems. The matrix elements I_{mn} are characteristic for the system. They are identical for all the systems in our macroscopic ensemble. Therefore in (B4) only the product $c_m c_n^*$ has to be averaged over the ensemble and we get

$$M_{xA} = N_o \gamma_A \sum_n \sum_m I_{nm} \overline{c_m c_n^*} \tag{B5}$$

where I_{nm} are the matrix elements of the operator I_{xA}. The only time variable elements in (B5) are the averaged products

$$\overline{c_m c_n^*}.$$

There are N^2 such products which, arranged in a square table, form the density matrix:

$$D = \begin{bmatrix} d_{11} & d_{12} & . & . & d_{1N} \\ d_{21} & d_{22} & . & . & d_{2N} \\ . & . & . & . & . \\ . & . & . & . & . \\ d_{N1} & d_{N2} & . & . & d_{NN} \end{bmatrix} \tag{B6}$$

with

$$d_{mn} = \overline{c_m c_n^*} \tag{B7}$$

We notice that $d_{nm} = d_{mn}^*$, i.e., D is a Hermitian matrix.

Density matrix and magnetizations

We rewrite (B5) making use of the expression (B7)

$$M_{xA} = N_o \gamma_A \sum_n \sum_m I_{nm} d_{mn} = N_o \gamma_A \sum_n \sum_m I_{nm} d_{nm}^* \tag{B8}$$

Relation (B8) represents the practical mode of calculating an observable magnetization component (in our case M_{xA}) when the density matrix D is known:

Multiply every matrix element of I_{xA} with the complex conjugate of the corresponding element of D and add all the products. Multiply the sum by $N_o \gamma_A \hbar$.

It is convenient to express the factor $N_o\gamma_A\hbar$ in terms of the equilibrium magnetization M_{oA}:

$$M_{oA} = \frac{N_o\gamma_A^2\hbar^2 I(I+1)B_o}{3kT} \tag{B9}$$

Note that M_{oA} is always a positive quantity, the absolute value of the equilibrium magnetization for nucleus A. For $I = 1/2$ the expression (B9) becomes:

$$M_{oA} = \frac{N_o\gamma_A^2\hbar^2 B_o}{4kT} \tag{B10}$$

In (I.3) we have introduced the quantity

$$p = \frac{h\nu_A}{kT} = \frac{\hbar\omega_A}{kT} = \frac{\hbar|\gamma_A|B_o}{kT} \tag{B11}$$

related to the Boltzmann factor of nucleus A. In accordance with our sign convention (negative γ) this can be rewritten as

$$p = -\frac{\hbar\gamma_A B_o}{kT} \tag{B12}$$

and (B10) becomes

$$M_{oA} = -\frac{N_o\gamma_A\hbar p}{4}$$

The factor $N_o\gamma_A\hbar$ in (B8) can now be written in the more convenient form

$$N_o\gamma_A\hbar = -\frac{4}{p}M_{oA} \tag{B13}$$

For nucleus X (see I.4) the factor is

$$N_o\gamma_X\hbar = -\frac{4}{q}M_{oX}$$

Let us apply the "recipe" for finding the magnetization components to the system AX (two spin 1/2 nuclei). The number of states is N=4 and the (Hermitian) density matrix has the rank 4:

$$D = \begin{bmatrix} d_{11} & d_{12} & d_{13} & d_{14} \\ d_{21} & d_{22} & d_{23} & d_{24} \\ d_{31} & d_{32} & d_{33} & d_{34} \\ d_{41} & d_{42} & d_{43} & d_{44} \end{bmatrix}$$

where $d_{kj} = d^*_{jk}$ and d_{jj} = real $(d_{jj} = d^*_{jj})$.

The angular momentum components for the AX system are given in (C12) through (C15). We have for instance

$$I_{zA} = \frac{1}{2}\begin{bmatrix} 1 & 0 & 0 & 0 \\ 0 & -1 & 0 & 0 \\ 0 & 0 & 1 & 0 \\ 0 & 0 & 0 & -1 \end{bmatrix} \quad \text{(B14)}$$

in which only four matrix elements out of 16 are nonvanishing.

Using (B8) and (B14) in this particular case we get:

$$M_{zA} = -\frac{4}{p} M_{oA}\left(\frac{d_{11}}{2} - \frac{d_{22}}{2} + \frac{d_{33}}{2} - \frac{d_{44}}{2}\right)$$

$$= -\frac{2}{p} M_{oA}\left(d_{11} - d_{22} + d_{33} - d_{44}\right) \quad \text{(B15)}$$

In the x direction

$$I_{xA} = \frac{1}{2}\begin{bmatrix} 0 & 1 & 0 & 0 \\ 1 & 0 & 0 & 0 \\ 0 & 0 & 0 & 1 \\ 0 & 0 & 1 & 0 \end{bmatrix} \quad \text{(B16)}$$

and the "recipe" leads to

$$M_{xA} = -\frac{2}{p} M_{oA}\left(d^*_{12} + d^*_{21} + d^*_{34} + d^*_{43}\right) \tag{B17}$$

In the same way we obtain

$$M_{yA} = \frac{2i}{p} M_{oA}\left(d^*_{12} - d^*_{21} + d^*_{34} - d^*_{43}\right) \tag{B18}$$

It is always convenient to combine M_x and M_y in one complex quantity, the transverse magnetization

$$M_T = M_x + iM_y \tag{B19}$$

This leads to the simplified form

$$M_{TA} = -\frac{4}{p} M_{oA}\left(d^*_{12} + d^*_{34}\right) \tag{B20}$$

The magnetization components for the other nucleus of the system, nucleus X, are given by

$$M_{zX} = -\frac{2}{q} M_{oX}\left(d_{11} + d_{22} - d_{33} - d_{44}\right) \tag{B21}$$

$$M_{TX} = -\frac{4}{q} M_{oX}\left(d^*_{13} + d^*_{24}\right) \tag{B22}$$

This is equivalent to calculating the trace (sum of diagonal elements) of the product $I \times D$:

$$\sum_n \sum_m I_{nm} d_{mn} = Tr(I \times D) \tag{B23}$$

The density matrix at thermal equilibrium

At equilibrium the nondiagonal elements are null because of the random phase distribution of the complex coefficients c_m. We denote with ϕ_m the phase of the complex quantity c_m:

$$c_m = |c_m| \exp(i\phi_m) \tag{B24}$$

A nondiagonal matrix element is

$$d_{mn} = \overline{c_m c_n^*} = \overline{|c_m| \cdot |c_n| \exp[i(\phi_m - \phi_n)]} \tag{B25}$$

The phase difference $\phi_m - \phi_n$ can have any value within 0 and 2π with equal probability. The complex number $\exp[i(\phi_m - \phi_n)]$, described as a vector in the complex plane, may be oriented in any direction. The average of a multitude of such vectors is null.

The diagonal elements are not null since $\phi_m - \phi_m = 0$.

$$d_{mm} = \overline{c_m c_m^*} = \overline{|c_m|^2} \tag{B26}$$

In quantum mechanics $|c_m|^2$ is the probablity of finding the system in the state $|m\rangle$; therefore $d_{mm} = P_m$ is the population of this state. The density matrix at equilibrium is

$$D = \begin{bmatrix} P_1 & 0 & . & . & 0 \\ 0 & P_2 & . & . & 0 \\ . & . & . & . & . \\ . & . & . & . & . \\ 0 & 0 & . & . & P_N \end{bmatrix} \tag{B27}$$

where $\sum P_n = 1$ (normalized populations).

Evolution of the density matrix between pulses

In the absence of the r.f. excitation the Hamiltonian $\boldsymbol{H}$ accepts the kets $|n\rangle$ as eigenfunctions:

$$\boldsymbol{H}|n\rangle = E_n|n\rangle \tag{B28}$$

and the Schrödinger equation

$$\frac{-\hbar}{i} \cdot \frac{\partial \varphi}{\partial t} = \boldsymbol{H}\varphi$$

becomes:

$$\frac{-\hbar}{i}\sum\frac{\mathrm{d}c_n}{\mathrm{d}t}\cdot|n\rangle=\sum c_n E_n|n\rangle \tag{B29}$$

Rearranging (B29) gives

$$\sum\left[\left(c_n E_n+\frac{\hbar}{i}\cdot\frac{\mathrm{d}c_n}{\mathrm{d}t}\right)\cdot|n\rangle\right]=0 \tag{B30}$$

Due to the orthogonality of the eigenfunctions, (B30) is satisfied only if each term of the sum is null:

$$c_n E_n+\frac{\hbar}{i}\cdot\frac{\mathrm{d}c_n}{\mathrm{d}t}=0 \tag{B31}$$

Hence

$$\frac{1}{c_n}\cdot\frac{\mathrm{d}c_n}{\mathrm{d}t}=\frac{-iE_n}{\hbar} \tag{B32}$$

or

$$\frac{\mathrm{d}}{\mathrm{d}t}(\ln c_n)=\frac{-iE_n}{\hbar} \tag{B33}$$

Integrating (B33) yields:

$$\ln c_n=-\frac{iE_n}{\hbar}t+C \tag{B34}$$

$$c_n=\exp\left(\frac{-iE_n}{\hbar}t\right)\cdot\exp(C) \tag{B35}$$

The integration constant C may be related to the value of c_n at time $t=0$. This gives $c_n(0)=\exp(C)$ and (B35) becomes

$$c_n=c_n(0)\cdot\exp\left(\frac{-iE_n}{\hbar}t\right) \tag{B36}$$

Knowing the evolution of all c_n coefficients will allow us to predict the time variation of the density matrix, hence that of the magnetization:

$$d_{mn} = \overline{c_m c_n^*} = \overline{c_m(0) \cdot c_n^*(0)} \cdot \exp\left(\frac{-i(E_m - E_n)}{\hbar} t \right)$$

$$= d_{mn}(0) \cdot \exp(-i\omega_{mn} t) \tag{B37}$$

We have demonstrated here the relation (I.13) used in the density matrix treatment to describe the evolution between pulses. The populations are invariable because $E_m - E_m = 0$ (relaxation processes are neglected throughout this book).

Effects of radiofrequency pulses

We have to find the time evolution of the density matrix under a given Hamiltonian, as we did in the previous section, but there are two things that make the problem more complicated.

First, the Hamiltonian is now time-dependent (radiofrequency magnetic field). This problem can be circumvented by describing the evolution of the system in a rotating frame, in which the rotating magnetic field appears as an immobile vector B_1, while the main magnetic field B_o is replaced by

$$\Delta B = B_o - \omega_{tr} / |\gamma|$$

The resultant of B_1 and ΔB is the effective field B_{eff} (Figure B.1).

The field B_1 is usually much larger than ΔB and the effective field practically is B_1. The Hamiltonian in the rotating frame is then

$$\boldsymbol{H} = |\gamma| \hbar B_1 I_x \tag{B38}$$

as we have assumed that B_1 is applied along the x axis of the rotating frame. For comparison, in the absence of the r.f. field, the Hamiltonian in the rotating frame is

$$\boldsymbol{H} = |\gamma| \hbar \Delta B I_z \tag{B39}$$

while in the laboratory frame it has the expression

$$\boldsymbol{H} = |\gamma| \hbar B_o I_z \tag{B40}$$

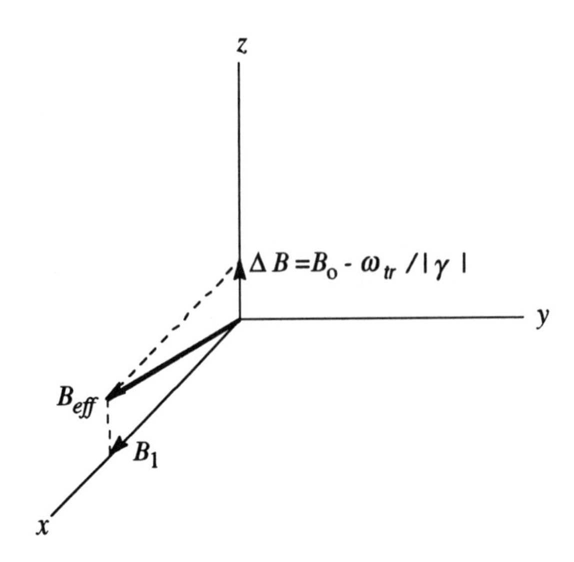

Figure B.1. The effective magnetic field B_{eff} in the rotating frame. *Ox*, *Oy* = axes of the rotating frame ; ω_{tr} = angular frequency of r.f. transmitter (angular velocity of the rotating frame)

Now we come to the second difficulty which prevents us from using the same approach as in the previous section: the new Hamiltonian (B38) does not have the kets $|n\rangle$ as eigenfunctions because we have passed from B_0 to B_1. We have to use a more general equation which describes the evolution of D under any Hamiltonian:

$$\frac{\mathrm{d}D}{\mathrm{d}t} = \frac{i}{\hbar}(DH - HD) \tag{B41}$$

the demonstration of which is given separately in the following section. The solution of (B41) is:

$$D(t) = \exp(-iHt/\hbar)D(0)\exp(iHt/\hbar) \tag{B42}$$

This can be verified by calculating the time derivative of (B42):

$$\frac{\mathrm{d}D}{\mathrm{d}t}=\left[\frac{\mathrm{d}}{\mathrm{d}t}\exp(-i\boldsymbol{H}t/\hbar)\right]D(0)\exp(i\boldsymbol{H}t/\hbar)$$

$$+\exp(-i\boldsymbol{H}t/\hbar)\left[\frac{\mathrm{d}D(0)}{\mathrm{d}t}\right]\exp(i\boldsymbol{H}t/\hbar)$$

$$+\exp(-i\boldsymbol{H}t/\hbar)D(0)\left[\frac{\mathrm{d}}{\mathrm{d}t}\exp(i\boldsymbol{H}t/\hbar)\right]$$

$$=\frac{-i\boldsymbol{H}}{\hbar}\exp(-i\boldsymbol{H}t/\hbar)D(0)\exp(i\boldsymbol{H}t/\hbar)+0$$

$$+\exp(-i\boldsymbol{H}t/\hbar)D(0)\frac{i\boldsymbol{H}}{\hbar}\exp(i\boldsymbol{H}t/\hbar)$$

$$=(-i\boldsymbol{H}/\hbar)D+D(i\boldsymbol{H}/\hbar)=i/\hbar(D\boldsymbol{H}-\boldsymbol{H}D)$$

In the particular case of a strong r.f. field B_1 applied along the x axis of the rotating frame we have, according to (B38):

$$i\boldsymbol{H}t/\hbar=i|\gamma|B_1I_xt=i\alpha I_x \qquad \text{(B43)}$$

where $\alpha=|\gamma|B_1t$ is the rotation angle of the magnetization around B_1 in the time t (pulse duration). Relation (B42) becomes

$$D=\exp(-iI_x\alpha)D(0)\exp(-iI_x\alpha)$$

$$=R^{-1}D(0)R \qquad \text{(B44)}$$

where R is the rotation operator [cf (A6)-(A7)]; $D(0)$ and D denote the density matrix before and after the pulse. We have thus demonstrated the relation (I.8).

In order to get an explicit matrix expression for R we have to calculate

$$R_{\alpha x}=\exp(i\alpha I_x) \qquad \text{(B45)}$$

using a series expansion of the exponential [see(A11)],

$$R_{\alpha x}=1+i\alpha I_x+\frac{(i\alpha)^2I_x{}^2}{2!}+\frac{(i\alpha)^3I_x{}^3}{3!}+\ldots \qquad \text{(B46)}$$

The powers of I_x may easily be calculated if one notices that

$$I_x^2 = \left(\frac{1}{4}\right) \cdot [\mathbf{1}] \tag{B47}$$

$$I_x = \frac{1}{2}\begin{bmatrix} 0 & 1 \\ 1 & 0 \end{bmatrix}$$

$$I_x^2 = \frac{1}{4}\begin{bmatrix} 1 & 0 \\ 0 & 1 \end{bmatrix} = \left(\frac{1}{4}\right)[\mathbf{1}]$$

$$I_x^3 = \frac{1}{4} I_x = \frac{1}{8}\begin{bmatrix} 0 & 1 \\ 1 & 0 \end{bmatrix} = \left(\frac{1}{8}\right) 2 I_x$$

$$I_x^4 = \frac{1}{16}\begin{bmatrix} 1 & 0 \\ 0 & 1 \end{bmatrix} = \left(\frac{1}{16}\right)[\mathbf{1}]$$

In general, for n = even we have:

$$I_x^n = \frac{1}{2^n}\begin{bmatrix} 1 & 0 \\ 0 & 1 \end{bmatrix} = \left(\frac{1}{2^n}\right) \cdot [\mathbf{1}] \tag{B48}$$

and for n = odd

$$I_x^n = \frac{1}{2^n}\begin{bmatrix} 0 & 1 \\ 1 & 0 \end{bmatrix} = \left(\frac{1}{2^n}\right) \cdot 2 I_x \tag{B49}$$

By using (C12) and (C14) one can verify that (B48) and (B49) are also true for I_{xA} and I_{xX} in a two spin system with I=1/2.

Introducing (B48) and (B49) in (B46) and separating the even and odd terms gives:

$$R_{\alpha x} = \left[1 - \frac{(\alpha/2)^2}{2!} + \frac{(\alpha/2)^4}{4!} - \ldots\right] \times \begin{bmatrix} 1 & 0 \\ 0 & 1 \end{bmatrix}$$

$$+ i\left[\alpha/2 - \frac{(\alpha/2)^3}{3!} + \frac{(\alpha/2)^5}{5!} - \ldots\right] \times \begin{bmatrix} 0 & 1 \\ 1 & 0 \end{bmatrix} \tag{B50}$$

Recognizing the sine (A12) and cosine (A13) series expansions we can write

$$R_{\alpha x} = \cos\frac{\alpha}{2}\cdot\begin{bmatrix}1 & 0\\ 0 & 1\end{bmatrix} + i\sin\frac{\alpha}{2}\cdot\begin{bmatrix}0 & 1\\ 1 & 0\end{bmatrix} = \cos\frac{\alpha}{2}\cdot[\mathbf{1}] + i\sin\frac{\alpha}{2}\cdot(2I_x)$$

$$= \begin{bmatrix}\cos\frac{\alpha}{2} & 0\\ 0 & \cos\frac{\alpha}{2}\end{bmatrix} + \begin{bmatrix}0 & i\sin\frac{\alpha}{2}\\ i\sin\frac{\alpha}{2} & 0\end{bmatrix} = \begin{bmatrix}\cos\frac{\alpha}{2} & i\sin\frac{\alpha}{2}\\ i\sin\frac{\alpha}{2} & \cos\frac{\alpha}{2}\end{bmatrix} \quad \text{(B51)}$$

For nucleus A in an AX system the rotation operator is

$$R_{\alpha x} = \begin{bmatrix}\cos\frac{\alpha}{2} & i\sin\frac{\alpha}{2} & 0 & 0\\ i\sin\frac{\alpha}{2} & \cos\frac{\alpha}{2} & 0 & 0\\ 0 & 0 & \cos\frac{\alpha}{2} & i\sin\frac{\alpha}{2}\\ 0 & 0 & i\sin\frac{\alpha}{2} & \cos\frac{\alpha}{2}\end{bmatrix} \quad \text{(B52)}$$

If the pulse is applied along the y axis, relations similar to (B48), (B49) apply:

$$I_y^n = \left(\frac{1}{2^n}\right)\cdot[\mathbf{1}] \qquad \text{for } n = \text{even}$$

$$I_y^n = \left(\frac{1}{2^n}\right)\cdot(2I_y) \qquad \text{for } n = \text{odd} \quad \text{(B53)}$$

and we obtain

$$R_{\alpha y} = \exp(i\alpha I_y) = \cos\frac{\alpha}{2}\cdot[\mathbf{1}] + i\sin\frac{\alpha}{2}\cdot(2I_y) \quad \text{(B54)}$$

Appendix C contains angular momentum components and rotation operators in matricial form, for a variety of spin systems and pulses. The reader may check some of those results by making $\alpha = 90^\circ$ or $\alpha = 180^\circ$ in the relations above.

If the radiofrequency field B_1 is applied along the $-x$ axis, it has the same effect as a pulse along the $+x$ axis, only the sense of rotation is reversed (left hand instead of right hand rule). The result of such a pulse is therefore a rotation by $-\alpha$ around Ox:

$$R_{\alpha(-x)} = R_{(-\alpha)x} = \exp(-i\alpha I_x) = \cos\frac{\alpha}{2}\cdot[\mathbf{1}] - i\sin\frac{\alpha}{2}\cdot(2I_x) \qquad \text{(B55)}$$

It is possible to extend the DM treatment to pulses with any phase (not only the four cardinal phases $x, y, -x, -y$) and/or off resonance pulses (B_{eff} does not coincide with B_1). We will not discuss them here because, as shown in the second part of the book, it is more convenient to handle them by means of the Product Operator formalism (see Appendix M).

Demonstratiom of (B41)

In order to demonstrate that

$$\frac{\mathrm{d}D}{\mathrm{d}t} = \frac{i}{\hbar}(DH - HD)$$

we follow the procedure used by Slichter (see Suggested Readings). We start with the (time dependent) Schrödinger equation

$$\frac{-\hbar}{i}\cdot\frac{\partial\varphi}{\partial t} = \boldsymbol{H}\varphi$$

where

$$\varphi = \sum_{n=1}^{N} c_n |n\rangle$$

with the observation that $|n\rangle$ are not assumed to be eigenfunctions of $\boldsymbol{H}$. Combining the last two equations gives

$$\frac{-\hbar}{i}\sum_{n=1}^{N}\frac{\mathrm{d}c_n}{\mathrm{d}t}\cdot|n\rangle = \sum_{n=1}^{N} c_n \boldsymbol{H}|n\rangle \qquad \text{(B56)}$$

If we premultiply this equation with the bra $\langle m|$ we get

$$\frac{-\hbar}{i}\sum_{n=1}^{N}\frac{\mathrm{d}c_n}{\mathrm{d}t}\langle m|n\rangle=\sum_{n=1}^{N}c_n\langle m|\boldsymbol{H}|n\rangle \tag{B57}$$

The choice of normalized and orthogonal functions for the basis set $|n\rangle$ implies

$$\begin{aligned}\langle m|n\rangle&=0 \qquad \text{for } m\neq n\\ \langle m|n\rangle&=1 \qquad \text{for } m=n\end{aligned} \tag{B58}$$

On the other hand $\langle m|H|n\rangle$ is the matrix element H_{mn} in the matrix representation of the Hamiltonian, so (B57) becomes

$$\frac{-\hbar}{i}\frac{\mathrm{d}c_n}{\mathrm{d}t}=\sum_{n=1}^{N}c_nH_{mn} \tag{B59}$$

If we consider now the product

$$p_{jk}=c_jc_k^* \tag{B60}$$

its time derivative will be

$$\frac{\mathrm{d}p_{jk}}{\mathrm{d}t}=\frac{\mathrm{d}c_j}{\mathrm{d}t}c_k^*+c_j\frac{\mathrm{d}c_k^*}{\mathrm{d}t}$$

$$=\frac{\mathrm{d}c_j}{\mathrm{d}t}c_k^*+c_j\left(\frac{\mathrm{d}c_k}{\mathrm{d}t}\right)^*$$

$$=\frac{-i}{\hbar}\left(\sum_{n=1}^{N}c_nH_{jn}\right)c_k^*+c_j\left[\frac{-i}{\hbar}\left(\sum_{n=1}^{N}c_nH_{kn}\right)\right]^*$$

$$=\frac{i}{\hbar}\left(\sum_{n=1}^{N}c_jc_n^*H_{kn}^*-\sum_{n=1}^{N}c_nc_k^*H_{jn}\right)$$

The change of sign comes from $(-i)^*=i$.

If we take into account that $\boldsymbol{H}$ is Hermitian ($H^*_{kn} = H_{nk}$) we get

$$\frac{\mathrm{d}p_{jk}}{\mathrm{d}t} = \frac{i}{\hbar}\left(\sum_{n=1}^{N} p_{jn}H_{nk} - \sum_{n=1}^{N} H_{jn}p_{nk}\right) \tag{B61}$$

The density matrix element d_{jk} is nothing other than the product p_{jk} averaged over the whole ensemble:

$$d_{jk} = \overline{p_{jk}}$$

On the other hand the Hamiltonian and its matrix elements are the same for all the systems within the ensemble, they are not affected by the operation of averaging. Taking the average on both sides of (B61) yields

$$\frac{\mathrm{d}d_{jk}}{\mathrm{d}t} = \frac{i}{\hbar}\left(\sum_{n=1}^{N} d_{jn}H_{nk} - \sum_{n=1}^{N} H_{jn}d_{nk}\right) \tag{B62}$$

According to the matrix multiplication rule (see Appendix A) the sums in (B62) represent matrix elements of the products $\boldsymbol{DH}$ and $\boldsymbol{HD}$, so (B62) can be written as

$$\left(\frac{\mathrm{d}\boldsymbol{D}}{\mathrm{d}t}\right)_{jk} = \frac{i}{\hbar}\left[(\boldsymbol{DH})_{jk} - (\boldsymbol{HD})_{jk}\right] = \frac{i}{\hbar}(\boldsymbol{DH} - \boldsymbol{HD})_{jk}$$

This demonstrates (B41) since the time derivative of a matrix is performed by taking the derivative of each element.

APPENDIX C: ANGULAR MOMENTUM AND ROTATION OPERATORS

System (spin): A(1/2)

$$I_x = \frac{1}{2}\begin{bmatrix} 0 & 1 \\ 1 & 0 \end{bmatrix} \quad ; \quad I_y = \frac{1}{2}\begin{bmatrix} 0 & -i \\ i & 0 \end{bmatrix} \tag{C1}$$

$$I_z = \frac{1}{2}\begin{bmatrix} 1 & 0 \\ 0 & -1 \end{bmatrix} \quad ; \quad I_x + iI_y = \begin{bmatrix} 0 & 1 \\ 0 & 0 \end{bmatrix} \tag{C2}$$

$$R_{\alpha x} = \begin{bmatrix} \cos\frac{\alpha}{2} & i\sin\frac{\alpha}{2} \\ i\sin\frac{\alpha}{2} & \cos\frac{\alpha}{2} \end{bmatrix} \quad ; \quad R_{\alpha y} = \begin{bmatrix} \cos\frac{\alpha}{2} & \sin\frac{\alpha}{2} \\ -\sin\frac{\alpha}{2} & \cos\frac{\alpha}{2} \end{bmatrix} \tag{C3}$$

$$R_{90x} = \frac{1}{\sqrt{2}}\begin{bmatrix} 1 & i \\ i & 1 \end{bmatrix} \quad ; \quad R_{90y} = \frac{1}{\sqrt{2}}\begin{bmatrix} 1 & 1 \\ -1 & 1 \end{bmatrix} \tag{C4}$$

$$R_{180x} = \begin{bmatrix} 0 & i \\ i & 0 \end{bmatrix} \quad ; \quad R_{180y} = \begin{bmatrix} 0 & 1 \\ -1 & 0 \end{bmatrix} \tag{C5}$$

System (spin): A(1)

$$I_x = \frac{1}{\sqrt{2}}\begin{bmatrix} 0 & 1 & 0 \\ 1 & 0 & 1 \\ 0 & 1 & 0 \end{bmatrix} \quad ; \quad I_y = \frac{1}{\sqrt{2}}\begin{bmatrix} 0 & -i & 0 \\ i & 0 & -i \\ 0 & i & 0 \end{bmatrix} \tag{C6}$$

$$I_z = \begin{bmatrix} 1 & 0 & 0 \\ 0 & 0 & 0 \\ 0 & 0 & -1 \end{bmatrix} \quad ; \quad I_x + iI_y = \sqrt{2}\begin{bmatrix} 0 & 1 & 0 \\ 0 & 0 & 1 \\ 0 & 0 & 0 \end{bmatrix} \tag{C7}$$

$$R_{\alpha x} = \frac{1}{2}\begin{bmatrix} \cos\alpha + 1 & i\sqrt{2}\sin\alpha & \cos\alpha - 1 \\ i\sqrt{2}\sin\alpha & 2\cos\alpha & i\sqrt{2}\sin\alpha \\ \cos\alpha - 1 & i\sqrt{2}\sin\alpha & \cos\alpha + 1 \end{bmatrix} \tag{C8}$$

$$R_{\alpha y} = \frac{1}{2}\begin{bmatrix} 1 + \cos\alpha & \sqrt{2}\sin\alpha & 1 - \cos\alpha \\ -\sqrt{2}\sin\alpha & 2\cos\alpha & \sqrt{2}\sin\alpha \\ 1 - \cos\alpha & -\sqrt{2}\sin\alpha & 1 + \cos\alpha \end{bmatrix} \tag{C9}$$

$$R_{90x} = \frac{1}{2}\begin{bmatrix} 1 & i\sqrt{2} & -1 \\ i\sqrt{2} & 0 & i\sqrt{2} \\ -1 & i\sqrt{2} & 1 \end{bmatrix} \; ; \; R_{90y} = \frac{1}{2}\begin{bmatrix} 1 & \sqrt{2} & 1 \\ -\sqrt{2} & 0 & \sqrt{2} \\ 1 & -\sqrt{2} & 1 \end{bmatrix} \tag{C10}$$

$$R_{180x} = \begin{bmatrix} 0 & 0 & -1 \\ 0 & -1 & 0 \\ -1 & 0 & 0 \end{bmatrix} \quad ; \quad R_{180y} = \begin{bmatrix} 0 & 0 & 1 \\ 0 & -1 & 0 \\ 1 & 0 & 1 \end{bmatrix} \tag{C11}$$

System (spin): A(1/2) X(1/2)

$$I_{xA} = \frac{1}{2}\begin{bmatrix} 0 & 1 & 0 & 0 \\ 1 & 0 & 0 & 0 \\ 0 & 0 & 0 & 1 \\ 0 & 0 & 1 & 0 \end{bmatrix} \quad ; \quad I_{yA} = \frac{1}{2}\begin{bmatrix} 0 & -i & 0 & 0 \\ i & 0 & 0 & 0 \\ 0 & 0 & 0 & -i \\ 0 & 0 & i & 0 \end{bmatrix} \tag{C12}$$

$$I_{zA} = \frac{1}{2}\begin{bmatrix} 1 & 0 & 0 & 0 \\ 0 & -1 & 0 & 0 \\ 0 & 0 & 1 & 0 \\ 0 & 0 & 0 & -1 \end{bmatrix} ;(I_x + iI_y)_A = \begin{bmatrix} 0 & 1 & 0 & 0 \\ 0 & 0 & 0 & 0 \\ 0 & 0 & 0 & 1 \\ 0 & 0 & 0 & 0 \end{bmatrix} \tag{C13}$$

$$I_{xX} = \frac{1}{2}\begin{bmatrix} 0 & 0 & 1 & 0 \\ 0 & 0 & 0 & 1 \\ 1 & 0 & 0 & 0 \\ 0 & 1 & 0 & 0 \end{bmatrix}; \qquad I_{yX} = \frac{1}{2}\begin{bmatrix} 0 & 0 & -i & 0 \\ 0 & 0 & 0 & -i \\ i & 0 & 0 & 0 \\ 0 & i & 0 & 0 \end{bmatrix} \tag{C14}$$

$$I_{zX} = \frac{1}{2}\begin{bmatrix} 1 & 0 & 0 & 0 \\ 0 & 1 & 0 & 0 \\ 0 & 0 & -1 & 0 \\ 0 & 0 & 0 & -1 \end{bmatrix}; (I_x + iI_y)_X = \begin{bmatrix} 0 & 0 & 1 & 0 \\ 0 & 0 & 0 & 1 \\ 0 & 0 & 0 & 0 \\ 0 & 0 & 0 & 0 \end{bmatrix} \tag{C15}$$

$$R_{90xA} = \frac{1}{\sqrt{2}}\begin{bmatrix} 1 & i & 0 & 0 \\ i & 1 & 0 & 0 \\ 0 & 0 & 1 & i \\ 0 & 0 & i & 1 \end{bmatrix}; \quad R_{90yA} = \frac{1}{\sqrt{2}}\begin{bmatrix} 1 & 1 & 0 & 0 \\ -1 & 1 & 0 & 0 \\ 0 & 0 & 1 & 1 \\ 0 & 0 & -1 & 1 \end{bmatrix} \tag{C16}$$

$$R_{180xA} = \begin{bmatrix} 0 & i & 0 & 0 \\ i & 0 & 0 & 0 \\ 0 & 0 & 0 & i \\ 0 & 0 & i & 0 \end{bmatrix}; \qquad R_{180yA} = \begin{bmatrix} 0 & 1 & 0 & 0 \\ -1 & 0 & 0 & 0 \\ 0 & 0 & 0 & 1 \\ 0 & 0 & -1 & 0 \end{bmatrix} \tag{C17}$$

$$R_{90xX} = \frac{1}{\sqrt{2}}\begin{bmatrix} 1 & 0 & i & 0 \\ 0 & 1 & 0 & i \\ i & 0 & 1 & 0 \\ 0 & i & 0 & 1 \end{bmatrix}; \quad R_{90yX} = \frac{1}{\sqrt{2}}\begin{bmatrix} 1 & 0 & 1 & 0 \\ 0 & 1 & 0 & 1 \\ -1 & 0 & 1 & 0 \\ 0 & -1 & 0 & 1 \end{bmatrix} \tag{C18}$$

$$R_{180xX} = \begin{bmatrix} 0 & 0 & i & 0 \\ 0 & 0 & 0 & i \\ i & 0 & 0 & 0 \\ 0 & i & 0 & 0 \end{bmatrix} \quad ; \quad R_{180yX} = \begin{bmatrix} 0 & 0 & 1 & 0 \\ 0 & 0 & 0 & 1 \\ -1 & 0 & 0 & 0 \\ 0 & -1 & 0 & 0 \end{bmatrix} \tag{C19}$$

Examples of *selective* rotation operators, affecting only one of the two possible transitions of nucleus X: 2-4 or 1-3.

$$R_{90x(24)} = \frac{1}{\sqrt{2}} \begin{bmatrix} \sqrt{2} & 0 & 0 & 0 \\ 0 & 1 & 0 & i \\ 0 & 0 & \sqrt{2} & 0 \\ 0 & i & 0 & 1 \end{bmatrix} ; R_{90y(24)} = \frac{1}{\sqrt{2}} \begin{bmatrix} \sqrt{2} & 0 & 0 & 0 \\ 0 & 1 & 0 & 1 \\ 0 & 0 & \sqrt{2} & 0 \\ 0 & -1 & 0 & 1 \end{bmatrix} \tag{C20}$$

$$R_{180x(24)} = \begin{bmatrix} 1 & 0 & 0 & 0 \\ 0 & 0 & 0 & i \\ 0 & 0 & 1 & 0 \\ 0 & i & 0 & 0 \end{bmatrix} \quad ; \quad R_{180y(24)} = \begin{bmatrix} 1 & 0 & 0 & 0 \\ 0 & 0 & 0 & 1 \\ 0 & 0 & 1 & 0 \\ 0 & -1 & 0 & 0 \end{bmatrix} \tag{C21}$$

$$R_{90x(13)} = \frac{1}{\sqrt{2}} \begin{bmatrix} 1 & 0 & i & 0 \\ 0 & \sqrt{2} & 0 & 0 \\ i & 0 & 1 & 0 \\ 0 & 0 & 0 & \sqrt{2} \end{bmatrix} ; R_{90y(13)} = \frac{1}{\sqrt{2}} \begin{bmatrix} 1 & 0 & 1 & 0 \\ 0 & \sqrt{2} & 0 & 0 \\ -1 & 0 & 1 & 0 \\ 0 & 0 & 0 & \sqrt{2} \end{bmatrix} \tag{C22}$$

$$R_{180x(13)} = \begin{bmatrix} 0 & 0 & i & 0 \\ 0 & 1 & 0 & 0 \\ i & 0 & 0 & 0 \\ 0 & 0 & 0 & 1 \end{bmatrix} \; ; \; R_{180y(13)} = \begin{bmatrix} 0 & 0 & 1 & 0 \\ 0 & 1 & 0 & 0 \\ -1 & 0 & 0 & 0 \\ 0 & 0 & 0 & 1 \end{bmatrix} \qquad \text{(C23)}$$

Selective rotation operators for the nucleus A (transition 1-2 or 3-4) can be written in a similar manner.

System (spin): A(1) X(1/2)

The energy states are labeled according to the figure below.

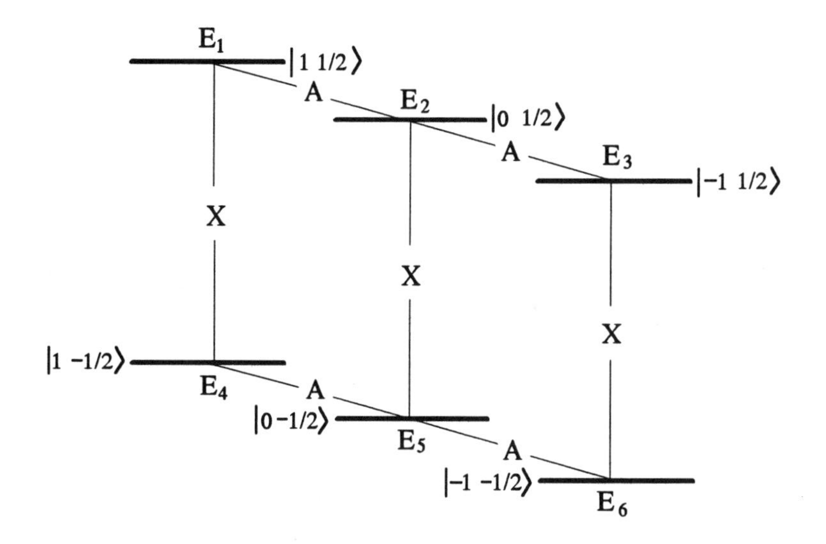

$$I_{xA} = \frac{1}{\sqrt{2}} \begin{bmatrix} 0 & 0 & 1 & 0 & 0 & 0 \\ 0 & 0 & 0 & 1 & 0 & 0 \\ 1 & 0 & 0 & 0 & 1 & 0 \\ 0 & 1 & 0 & 0 & 0 & 1 \\ 0 & 0 & 1 & 0 & 0 & 0 \\ 0 & 0 & 0 & 1 & 0 & 0 \end{bmatrix} \qquad \text{(C24)}$$

$$I_{xX} = \frac{1}{2}\begin{bmatrix} 0 & 1 & 0 & 0 & 0 & 0 \\ 1 & 0 & 0 & 0 & 0 & 0 \\ 0 & 0 & 0 & 1 & 0 & 0 \\ 0 & 0 & 1 & 0 & 0 & 0 \\ 0 & 0 & 0 & 0 & 0 & 1 \\ 0 & 0 & 0 & 0 & 1 & 0 \end{bmatrix} \tag{C25}$$

I_y can be written in the same way, taking (C6) and (C1) as starting points.

Examples of rotation operators for the AX(1, 1/2) system:

$$R_{90yA} = \frac{1}{2}\begin{bmatrix} 1 & 0 & \sqrt{2} & 0 & 1 & 0 \\ 0 & 1 & 0 & \sqrt{2} & 0 & 1 \\ -\sqrt{2} & 0 & 1 & 0 & \sqrt{2} & 0 \\ 0 & -\sqrt{2} & 0 & 1 & 0 & \sqrt{2} \\ 1 & 0 & -\sqrt{2} & 0 & 1 & 0 \\ 0 & 1 & 0 & -\sqrt{2} & 0 & 1 \end{bmatrix} \tag{C26}$$

$$R_{180xX} = \begin{bmatrix} 0 & i & 0 & 0 & 0 & 0 \\ i & 0 & 0 & 0 & 0 & 0 \\ 0 & 0 & 0 & i & 0 & 0 \\ 0 & 0 & i & 0 & 0 & 0 \\ 0 & 0 & 0 & 0 & 0 & i \\ 0 & 0 & 0 & 0 & i & 0 \end{bmatrix} \tag{C27}$$

Reciprocals R^{-1} of all rotation operators can be found through transposition and complex conjugation (see Appendix A).

Rotations about the z-axis

These rotation operators are needed in the following section (Phase Cycling). They can be derived in the same way as the x and y rotation operators [see (B45) to (B54)], after observing that

$$I_z^n = \left(\frac{1}{2^n}\right)\cdot[\mathbf{1}] \qquad \text{for n = even}$$

$$I_z^n = \left(\frac{1}{2^n}\right)\cdot(2I_z) \qquad \text{for n = odd} \tag{C28}$$

For the one spin system A(1/2) we have

$$R_{\alpha z} = \begin{bmatrix} b & 0 \\ 0 & b* \end{bmatrix} \tag{C29}$$

with
$$b = \cos(\alpha/2) + i\sin(\alpha/2) = \exp(i\alpha/2) \tag{C30}$$

For the two spin system A(1/2)X(1/2) we have

$$R_{\alpha zA} = \begin{bmatrix} b & 0 & 0 & 0 \\ 0 & b* & 0 & 0 \\ 0 & 0 & b & 0 \\ 0 & 0 & 0 & b* \end{bmatrix} \; ; \; R_{\alpha zX} = \begin{bmatrix} b & 0 & 0 & 0 \\ 0 & b & 0 & 0 \\ 0 & 0 & b* & 0 \\ 0 & 0 & 0 & b* \end{bmatrix} \tag{C31}$$

$$R_{\alpha zAX} = R_{\alpha zA} R_{\alpha zX} = \begin{bmatrix} b^2 & 0 & 0 & 0 \\ 0 & 1 & 0 & 0 \\ 0 & 0 & 1 & 0 \\ 0 & 0 & 0 & b*^2 \end{bmatrix} \tag{C32}$$

Phase cycling

This section contains rotation operators for phase cycled pulses. The rotation axis for such pulses is situated in the xy plane and makes an angle Φ with the x-axis. In succesive runs, the angle Φ assumes different values. When Φ is equal to 0°, 90°, 180°, or 270°, the rotation axis is x, y, $-x$, or $-y$, respectively. The expressions given in this section are valid for any value of Φ, even if it is not a multiple of 90°. Such values are seldom used in pulse sequences but they may be used to assess the effect of imperfect phases.

The rotation operator $R_{90\Phi}$ represents a 90° rotation about an axis making the angle Φ with the x-axis. In order to find the expression of $R_{90\Phi}$, we observe that this rotation is equivalent with the following succession of rotations:

a. A rotation by $-\Phi$ (clockwise) about Oz, bringing the rotation axis in line with Ox.
b. A 90° rotation about Ox.
c. A rotation by Φ (counterclockwise) about Oz.

For the one-spin system A(1/2), using (C29), this leads to

$$R_{90\Phi} = R_{(-\Phi)z} R_{90x} R_{\Phi z}$$

$$= \begin{bmatrix} b* & 0 \\ 0 & b \end{bmatrix} \frac{1}{\sqrt{2}} \begin{bmatrix} 1 & i \\ i & 1 \end{bmatrix} \begin{bmatrix} b & 0 \\ 0 & b* \end{bmatrix}$$

$$= \frac{1}{\sqrt{2}} \begin{bmatrix} b* & 0 \\ 0 & b \end{bmatrix} \begin{bmatrix} b & ib* \\ ib & b* \end{bmatrix} = \frac{1}{\sqrt{2}} \begin{bmatrix} 1 & ib*^2 \\ ib^2 & 1 \end{bmatrix} \tag{C33}$$

where $b = \exp(i\Phi / 2)$ from (C30). With the new notation

$$a = ib*^2 = i\exp(-i\Phi) \tag{C34}$$

the relation (C33) becomes

$$R_{90\Phi} = \frac{1}{\sqrt{2}} \begin{bmatrix} 1 & a \\ -a* & 1 \end{bmatrix} \tag{C35}$$

In a similar way one can demonstrate that

$$R_{180\Phi} = \begin{bmatrix} 0 & a \\ -a^* & 0 \end{bmatrix} \tag{C36}$$

For the two-spin system A(1/2)X(1/2)

$$R_{90\Phi A} = \frac{1}{\sqrt{2}} \begin{bmatrix} 1 & a & 0 & 0 \\ -a^* & 1 & 0 & 0 \\ 0 & 0 & 1 & a \\ 0 & 0 & -a^* & 1 \end{bmatrix} \tag{C37}$$

$$R_{90\Phi X} = \frac{1}{\sqrt{2}} \begin{bmatrix} 1 & 0 & a & 0 \\ 0 & 1 & 0 & a \\ -a^* & 0 & 1 & 0 \\ 0 & -a^* & 0 & 1 \end{bmatrix} \tag{C38}$$

$$R_{90\Phi AX} = R_{90\Phi A} R_{90\Phi X} = \frac{1}{2} \begin{bmatrix} 1 & a & a & a^2 \\ -a^* & 1 & -1 & a \\ -a^* & -1 & 1 & a \\ a^{*2} & -a^* & -a^* & 1 \end{bmatrix} \tag{C39}$$

This operator has been used in the DM treatment of INADEQUATE [see (I.83)].

One can verify that for $\Phi = 0$ we have $a = i$ and

$$R_{90\Phi AX} = R_{90xAX} \qquad \text{[cf.(I.34)]}$$

When $\Phi = 90^\circ$ we have $a = 1$ and

$$R_{90\Phi AX} = R_{90yAX} \qquad \text{[cf.(I.101)]}$$

For the 180° pulse, similar calculations lead to

$$R_{180\Phi A} = \begin{bmatrix} 0 & a & 0 & 0 \\ -a^* & 0 & 0 & 0 \\ 0 & 0 & 0 & a \\ 0 & 0 & -a^* & 0 \end{bmatrix} \qquad \text{(C40)}$$

$$R_{180\Phi X} = \begin{bmatrix} 0 & 0 & a & 0 \\ 0 & 0 & 0 & a \\ -a^* & 0 & 0 & 0 \\ 0 & -a^* & 0 & 0 \end{bmatrix} \qquad \text{(C41)}$$

$$R_{180\Phi AX} = \begin{bmatrix} 0 & 0 & 0 & a^2 \\ 0 & 0 & -1 & 0 \\ 0 & -1 & 0 & 0 \\ a^{*2} & 0 & 0 & 0 \end{bmatrix} \qquad \text{(C42)}$$

The 180° operators can be calculated by multiplying the respective 90° operator with itself (two successive 90° rotations).

Cyclops

Even in the simplest one-dimensional sequences, involving one single pulse (the "observe" pulse), a form of phase cycling is used in order to eliminate the radiofrequency interferences. The observe pulse is cycled through all four phases, e.g. clockwise: +x, – y, – x, +y. The f.i.d. phase follows the same pattern. There will be no accumulation unless the reciever phase is also cycled clockwise. An extraneous signal is not phase cycled and it will be averaged out because of the receiver cycling, provided the number of transients is a multiple of four. The procedure is known as "cyclops".

APPENDIX D: PROPERTIES OF PRODUCT OPERATORS

A system of m spin 1/2 nuclei has $N = 2^m$ states. The basis set for this system consists of N^2 product operators (PO) which are $N \times N$ hermitian matrices. We summarize here the most significant features of these matrices.

1. There is only one nonvanishing element per row. As a consequence, any PO has only N nonvanishing elements out of N^2 elements.

2. There is also only one nonvanishing element per column. Properties 1 and 2 are found in the matrices representing angular momentum components I_x, I_y, I_z (see Appendix C). A product of two matrices having these properties inherits them.

3. The nonvanishing elements of a PO are either ± 1 or $\pm i$.

4. If P_j and P_k are two product operators from the basis set, the trace (sum of diagonal elements) of their product is

$$\mathrm{Tr}(P_j P_k) = N\delta_{jk} \qquad \text{(D1)}$$

where d_{jk} (the Kronecker delta) has the value

$$d_{jk} = 0 \quad \text{if } j \neq k$$
$$d_{jk} = 1 \quad \text{if } j = k$$

The property (D1) illustrates the *orthogonality* of the PO's. The product of two different PO's is traceless. The square of a given PO is equal to the unit matrix, therefore its trace is equal to N.

Expressing a given matrix in terms of PO's

Since the basis set is a complete set, any $N \times N$ matrix can be expressed as a linear combination of PO's :

$$D = c_1 P_1 + c_2 P_2 + \ldots + c_L P_L \qquad \text{where} \qquad L = N^2 \qquad \text{(D2)}$$

Given the matrix D, the coefficients c_j can be determined using the orthogonality relation (D1).

$$\mathrm{Tr}(DP_j) = \sum_{k=1}^{m} c_k \mathrm{Tr}(P_k P_j) = \sum_{k=1}^{m} c_k N \delta_{kj} = N c_j \tag{D3}$$

Therefore

$$c_j = \frac{1}{N} \mathrm{Tr}(DP_j) \tag{D4}$$

In the PO treatment of NMR sequences we do not have to go through the routine described above since we start with the density matrix expressed in terms of POs and we have rules for any rotation or evolution which give the new density matrix also expressed in terms of POs. For the same reason, we do not need to know the POs in their matrix form in order to operate with them.

The complete basis set for $m = 2$ ($N = 4$) is given in Table II.1. We give in the following pages a few examples of POs in matrix form for $m = 3$ ($N = 8$) and for $m = 4$ ($N = 16$). At the end of this appendix a computer program can be found (written in BASIC) which will help generate all the product operators for $n = 2, 3$, or 4.

In all the matrices given below as examples, the dots represent zeros.

$$[111] = \begin{bmatrix} 1 & \cdot & \cdot & \cdot & \cdot & \cdot & \cdot & \cdot \\ \cdot & 1 & \cdot & \cdot & \cdot & \cdot & \cdot & \cdot \\ \cdot & \cdot & 1 & \cdot & \cdot & \cdot & \cdot & \cdot \\ \cdot & \cdot & \cdot & 1 & \cdot & \cdot & \cdot & \cdot \\ \cdot & \cdot & \cdot & \cdot & 1 & \cdot & \cdot & \cdot \\ \cdot & \cdot & \cdot & \cdot & \cdot & 1 & \cdot & \cdot \\ \cdot & \cdot & \cdot & \cdot & \cdot & \cdot & 1 & \cdot \\ \cdot & \cdot & \cdot & \cdot & \cdot & \cdot & \cdot & 1 \end{bmatrix} \qquad [11y] = \begin{bmatrix} \cdot & \cdot & \cdot & \cdot & -i & \cdot & \cdot & \cdot \\ \cdot & \cdot & \cdot & \cdot & \cdot & -i & \cdot & \cdot \\ \cdot & \cdot & \cdot & \cdot & \cdot & \cdot & -i & \cdot \\ \cdot & \cdot & \cdot & \cdot & \cdot & \cdot & \cdot & -i \\ i & \cdot & \cdot & \cdot & \cdot & \cdot & \cdot & \cdot \\ \cdot & i & \cdot & \cdot & \cdot & \cdot & \cdot & \cdot \\ \cdot & \cdot & i & \cdot & \cdot & \cdot & \cdot & \cdot \\ \cdot & \cdot & \cdot & i & \cdot & \cdot & \cdot & \cdot \end{bmatrix}$$

$$[z11] = \begin{bmatrix} 1 & \cdot & \cdot & \cdot & \cdot & \cdot & \cdot & \cdot \\ \cdot & -1 & \cdot & \cdot & \cdot & \cdot & \cdot & \cdot \\ \cdot & \cdot & 1 & \cdot & \cdot & \cdot & \cdot & \cdot \\ \cdot & \cdot & \cdot & -1 & \cdot & \cdot & \cdot & \cdot \\ \cdot & \cdot & \cdot & \cdot & 1 & \cdot & \cdot & \cdot \\ \cdot & \cdot & \cdot & \cdot & \cdot & -1 & \cdot & \cdot \\ \cdot & \cdot & \cdot & \cdot & \cdot & \cdot & 1 & \cdot \\ \cdot & \cdot & \cdot & \cdot & \cdot & \cdot & \cdot & -1 \end{bmatrix}$$

$$[1y1] = \begin{bmatrix} \cdot & \cdot & -i & \cdot & \cdot & \cdot & \cdot & \cdot \\ \cdot & \cdot & \cdot & -i & \cdot & \cdot & \cdot & \cdot \\ i & \cdot & \cdot & \cdot & \cdot & \cdot & \cdot & \cdot \\ \cdot & i & \cdot & \cdot & \cdot & \cdot & \cdot & \cdot \\ \cdot & \cdot & \cdot & \cdot & \cdot & \cdot & -i & \cdot \\ \cdot & \cdot & \cdot & \cdot & \cdot & \cdot & \cdot & -i \\ \cdot & \cdot & \cdot & \cdot & i & \cdot & \cdot & \cdot \\ \cdot & \cdot & \cdot & \cdot & \cdot & i & \cdot & \cdot \end{bmatrix}$$

$$[1xx] = \begin{bmatrix} \cdot & \cdot & \cdot & \cdot & \cdot & \cdot & 1 & \cdot \\ \cdot & \cdot & \cdot & \cdot & \cdot & \cdot & \cdot & 1 \\ \cdot & \cdot & \cdot & \cdot & 1 & \cdot & \cdot & \cdot \\ \cdot & \cdot & \cdot & \cdot & \cdot & 1 & \cdot & \cdot \\ \cdot & \cdot & 1 & \cdot & \cdot & \cdot & \cdot & \cdot \\ \cdot & \cdot & \cdot & 1 & \cdot & \cdot & \cdot & \cdot \\ 1 & \cdot & \cdot & \cdot & \cdot & \cdot & \cdot & \cdot \\ \cdot & 1 & \cdot & \cdot & \cdot & \cdot & \cdot & \cdot \end{bmatrix}$$

$$[xy1] = \begin{bmatrix} \cdot & \cdot & \cdot & -i & \cdot & \cdot & \cdot & \cdot \\ \cdot & \cdot & -i & \cdot & \cdot & \cdot & \cdot & \cdot \\ \cdot & i & \cdot & \cdot & \cdot & \cdot & \cdot & \cdot \\ i & \cdot & \cdot & \cdot & \cdot & \cdot & \cdot & \cdot \\ \cdot & \cdot & \cdot & \cdot & \cdot & \cdot & \cdot & -i \\ \cdot & \cdot & \cdot & \cdot & \cdot & \cdot & -i & \cdot \\ \cdot & \cdot & \cdot & \cdot & \cdot & i & \cdot & \cdot \\ \cdot & \cdot & \cdot & \cdot & i & \cdot & \cdot & \cdot \end{bmatrix}$$

$$[1xy] = \begin{bmatrix} \cdot & \cdot & \cdot & \cdot & \cdot & \cdot & -i & \cdot \\ \cdot & \cdot & \cdot & \cdot & \cdot & \cdot & \cdot & -i \\ \cdot & \cdot & \cdot & \cdot & -i & \cdot & \cdot & \cdot \\ \cdot & \cdot & \cdot & \cdot & \cdot & -i & \cdot & \cdot \\ \cdot & \cdot & i & \cdot & \cdot & \cdot & \cdot & \cdot \\ \cdot & \cdot & \cdot & i & \cdot & \cdot & \cdot & \cdot \\ i & \cdot & \cdot & \cdot & \cdot & \cdot & \cdot & \cdot \\ \cdot & i & \cdot & \cdot & \cdot & \cdot & \cdot & \cdot \end{bmatrix}$$

$$[xyz] = \begin{bmatrix} \cdot & \cdot & \cdot & -i & \cdot & \cdot & \cdot & \cdot \\ \cdot & \cdot & -i & \cdot & \cdot & \cdot & \cdot & \cdot \\ \cdot & i & \cdot & \cdot & \cdot & \cdot & \cdot & \cdot \\ i & \cdot & \cdot & \cdot & \cdot & \cdot & \cdot & \cdot \\ \cdot & \cdot & \cdot & \cdot & \cdot & \cdot & \cdot & i \\ \cdot & \cdot & \cdot & \cdot & \cdot & \cdot & i & \cdot \\ \cdot & \cdot & \cdot & \cdot & \cdot & -i & \cdot & \cdot \\ \cdot & \cdot & \cdot & \cdot & -i & \cdot & \cdot & \cdot \end{bmatrix}$$

$$[zxy] = \begin{bmatrix} \cdot & \cdot & \cdot & \cdot & \cdot & \cdot & -i & \cdot \\ \cdot & \cdot & \cdot & \cdot & \cdot & \cdot & \cdot & i \\ \cdot & \cdot & \cdot & \cdot & -i & \cdot & \cdot & \cdot \\ \cdot & \cdot & \cdot & \cdot & \cdot & i & \cdot & \cdot \\ \cdot & \cdot & i & \cdot & \cdot & \cdot & \cdot & \cdot \\ \cdot & \cdot & \cdot & -i & \cdot & \cdot & \cdot & \cdot \\ i & \cdot & \cdot & \cdot & \cdot & \cdot & \cdot & \cdot \\ \cdot & -i & \cdot & \cdot & \cdot & \cdot & \cdot & \cdot \end{bmatrix}$$

$$[x1y] = \begin{bmatrix} \cdot & \cdot & \cdot & \cdot & \cdot & -i & \cdot & \cdot \\ \cdot & \cdot & \cdot & \cdot & -i & \cdot & \cdot & \cdot \\ \cdot & \cdot & \cdot & \cdot & \cdot & \cdot & \cdot & -i \\ \cdot & \cdot & \cdot & \cdot & \cdot & \cdot & -i & \cdot \\ \cdot & i & \cdot & \cdot & \cdot & \cdot & \cdot & \cdot \\ i & \cdot & \cdot & \cdot & \cdot & \cdot & \cdot & \cdot \\ \cdot & \cdot & \cdot & i & \cdot & \cdot & \cdot & \cdot \\ \cdot & \cdot & i & \cdot & \cdot & \cdot & \cdot & \cdot \end{bmatrix}$$

$$[1zy] = \begin{bmatrix} \cdot & \cdot & \cdot & \cdot & -i & \cdot & \cdot & \cdot \\ \cdot & \cdot & \cdot & \cdot & \cdot & -i & \cdot & \cdot \\ \cdot & \cdot & \cdot & \cdot & \cdot & \cdot & i & \cdot \\ \cdot & \cdot & \cdot & \cdot & \cdot & \cdot & \cdot & i \\ i & \cdot & \cdot & \cdot & \cdot & \cdot & \cdot & \cdot \\ \cdot & i & \cdot & \cdot & \cdot & \cdot & \cdot & \cdot \\ \cdot & \cdot & -i & \cdot & \cdot & \cdot & \cdot & \cdot \\ \cdot & \cdot & \cdot & -i & \cdot & \cdot & \cdot & \cdot \end{bmatrix} \qquad [zzy] = \begin{bmatrix} \cdot & \cdot & \cdot & \cdot & -i & \cdot & \cdot & \cdot \\ \cdot & \cdot & \cdot & \cdot & \cdot & i & \cdot & \cdot \\ \cdot & \cdot & \cdot & \cdot & \cdot & \cdot & i & \cdot \\ \cdot & \cdot & \cdot & \cdot & \cdot & \cdot & \cdot & -i \\ i & \cdot & \cdot & \cdot & \cdot & \cdot & \cdot & \cdot \\ \cdot & -i & \cdot & \cdot & \cdot & \cdot & \cdot & \cdot \\ \cdot & \cdot & -i & \cdot & \cdot & \cdot & \cdot & \cdot \\ \cdot & \cdot & \cdot & i & \cdot & \cdot & \cdot & \cdot \end{bmatrix}$$

$$[yxy] = \begin{bmatrix} \cdot & \cdot & \cdot & \cdot & \cdot & \cdot & \cdot & -1 \\ \cdot & \cdot & \cdot & \cdot & \cdot & \cdot & 1 & \cdot \\ \cdot & \cdot & \cdot & \cdot & \cdot & -1 & \cdot & \cdot \\ \cdot & \cdot & \cdot & \cdot & 1 & \cdot & \cdot & \cdot \\ \cdot & \cdot & \cdot & 1 & \cdot & \cdot & \cdot & \cdot \\ \cdot & \cdot & -1 & \cdot & \cdot & \cdot & \cdot & \cdot \\ \cdot & 1 & \cdot & \cdot & \cdot & \cdot & \cdot & \cdot \\ -1 & \cdot & \cdot & \cdot & \cdot & \cdot & \cdot & \cdot \end{bmatrix} \qquad [yyy] = \begin{bmatrix} \cdot & \cdot & \cdot & \cdot & \cdot & \cdot & \cdot & i \\ \cdot & \cdot & \cdot & \cdot & \cdot & \cdot & -i & \cdot \\ \cdot & \cdot & \cdot & \cdot & \cdot & -i & \cdot & \cdot \\ \cdot & \cdot & \cdot & \cdot & i & \cdot & \cdot & \cdot \\ \cdot & \cdot & \cdot & -i & \cdot & \cdot & \cdot & \cdot \\ \cdot & \cdot & i & \cdot & \cdot & \cdot & \cdot & \cdot \\ \cdot & i & \cdot & \cdot & \cdot & \cdot & \cdot & \cdot \\ -i & \cdot & \cdot & \cdot & \cdot & \cdot & \cdot & \cdot \end{bmatrix}$$

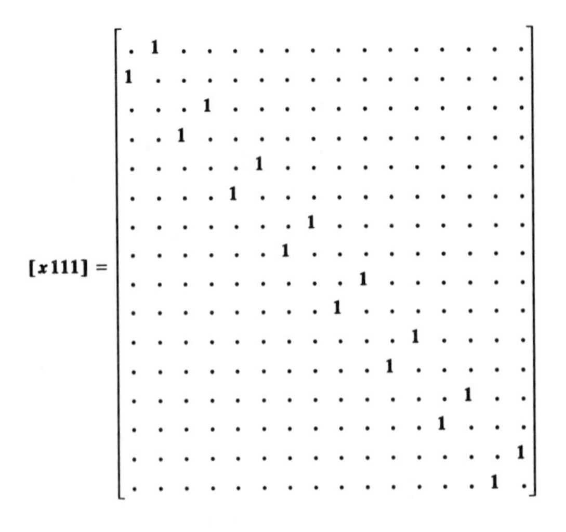

$$[1xx1] = \begin{bmatrix}
\cdot & \cdot & \cdot & \cdot & \cdot & \cdot & 1 & \cdot & \cdot & \cdot & \cdot & \cdot & \cdot & \cdot & \cdot & \cdot \\
\cdot & \cdot & \cdot & \cdot & \cdot & \cdot & \cdot & 1 & \cdot & \cdot & \cdot & \cdot & \cdot & \cdot & \cdot & \cdot \\
\cdot & \cdot & \cdot & \cdot & 1 & \cdot & \cdot & \cdot & \cdot & \cdot & \cdot & \cdot & \cdot & \cdot & \cdot & \cdot \\
\cdot & \cdot & \cdot & \cdot & \cdot & 1 & \cdot & \cdot & \cdot & \cdot & \cdot & \cdot & \cdot & \cdot & \cdot & \cdot \\
\cdot & \cdot & 1 & \cdot & \cdot & \cdot & \cdot & \cdot & \cdot & \cdot & \cdot & \cdot & \cdot & \cdot & \cdot & \cdot \\
\cdot & \cdot & \cdot & 1 & \cdot & \cdot & \cdot & \cdot & \cdot & \cdot & \cdot & \cdot & \cdot & \cdot & \cdot & \cdot \\
1 & \cdot & \cdot & \cdot & \cdot & \cdot & \cdot & \cdot & \cdot & \cdot & \cdot & \cdot & \cdot & \cdot & \cdot & \cdot \\
\cdot & 1 & \cdot & \cdot & \cdot & \cdot & \cdot & \cdot & \cdot & \cdot & \cdot & \cdot & \cdot & \cdot & \cdot & \cdot \\
\cdot & \cdot & \cdot & \cdot & \cdot & \cdot & \cdot & \cdot & \cdot & \cdot & \cdot & \cdot & \cdot & \cdot & 1 & \cdot \\
\cdot & \cdot & \cdot & \cdot & \cdot & \cdot & \cdot & \cdot & \cdot & \cdot & \cdot & \cdot & \cdot & \cdot & \cdot & 1 \\
\cdot & \cdot & \cdot & \cdot & \cdot & \cdot & \cdot & \cdot & \cdot & \cdot & \cdot & \cdot & 1 & \cdot & \cdot & \cdot \\
\cdot & \cdot & \cdot & \cdot & \cdot & \cdot & \cdot & \cdot & \cdot & \cdot & \cdot & \cdot & \cdot & 1 & \cdot & \cdot \\
\cdot & \cdot & \cdot & \cdot & \cdot & \cdot & \cdot & \cdot & \cdot & \cdot & 1 & \cdot & \cdot & \cdot & \cdot & \cdot \\
\cdot & \cdot & \cdot & \cdot & \cdot & \cdot & \cdot & \cdot & \cdot & \cdot & \cdot & 1 & \cdot & \cdot & \cdot & \cdot \\
\cdot & \cdot & \cdot & \cdot & \cdot & \cdot & \cdot & \cdot & 1 & \cdot & \cdot & \cdot & \cdot & \cdot & \cdot & \cdot \\
\cdot & \cdot & \cdot & \cdot & \cdot & \cdot & \cdot & \cdot & \cdot & 1 & \cdot & \cdot & \cdot & \cdot & \cdot & \cdot
\end{bmatrix}$$

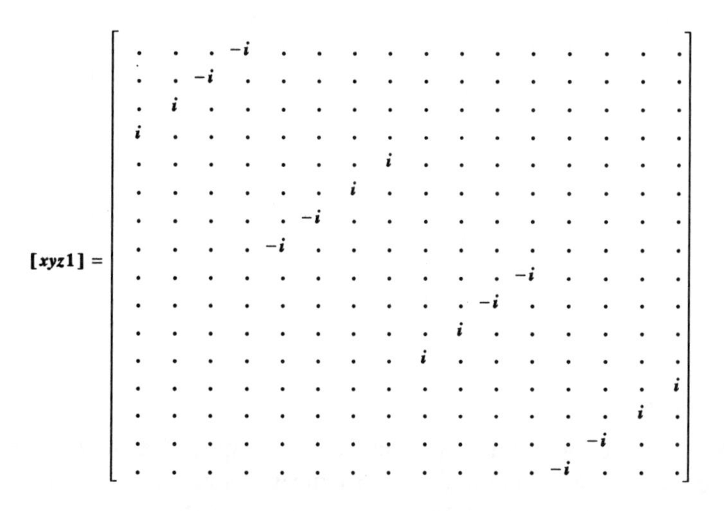

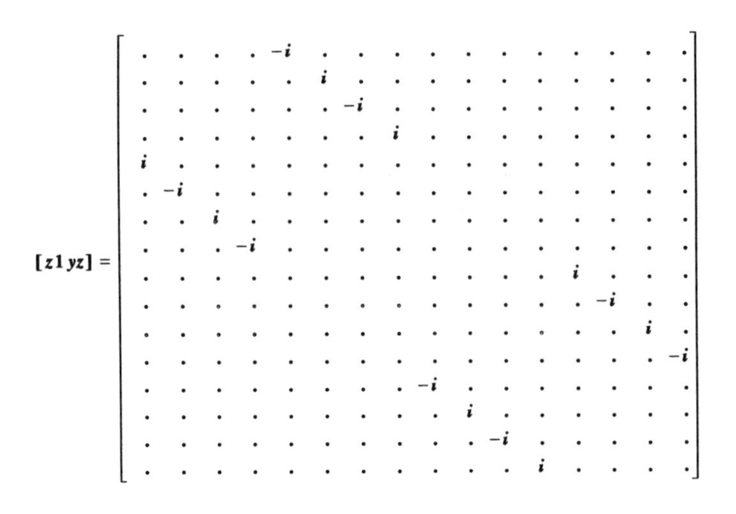

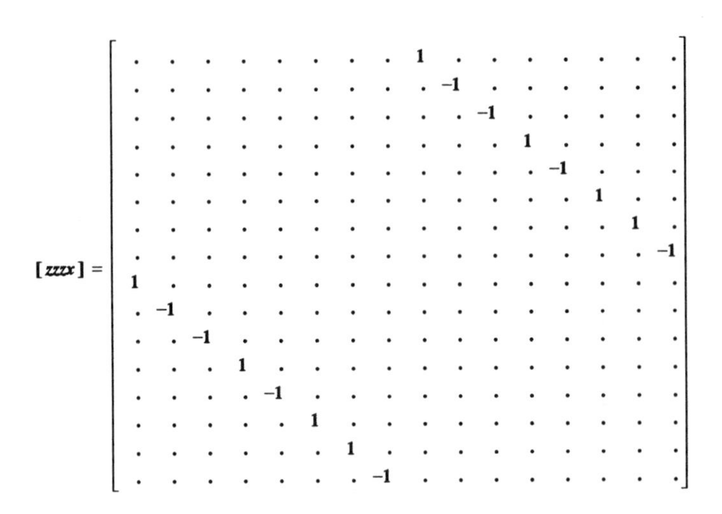

All the product operators sampled in this appendix have been calculated with the program POP (Product OPerators) listed on the following pages. It is written in BASIC, Version CPM-86, Rev.5.22 by Microsoft.

```
10  REM- PROGRAM "POP" CALCULATES AND PRINTS THE
11  REM- BASIC OPERATORS ACCORDING TO SORENSEN-
12  REM- ERNST FOR A SYSTEM OF TWO, THREE, OR FOUR
13  REM- SPIN 1/2 NUCLEI
20  DIM D%(256,16),N%(256,16),F%(256),J%(4),J$(4)
25  DIM B$(256),A(16,16),B(16,16)
30  PRINT "NUMBER OF SPIN 1/2 NUCLEI (2,3,OR 4)"
40  INPUT NN%
45  IF NN%>4 THEN NN%=4
46  IF NN%<2 THEN NN%=2
50  REM- NUMBER OF STATES
60  NS%=2^NN%
65  IF NN%=2 THEN B$="14 seconds"
66  IF NN%=3 THEN B$="38 seconds"
67  IF NN%=4 THEN B$="2 min 46 s"
68  PRINT "Please wait ";B$
70  GOSUB 940
71  REM- All product operators are now calculated and labeled
75  PRINT "                    M E N U"
76  PRINT "                    ========="
78  PRINT " A - Display a specified product operator"
80  PRINT " B - Display all product operators"
82  PRINT " C - Print a specified product operator (hard copy)"
84  PRINT " D - Print all product operators (hard copy)"
86  PRINT " E - Express a matrix in terms of product operators"
88  PRINT " F - Same as E but printed (hard copy)"
118 PRINT " X - Exit MENU"
120 M%=14
125 FOR I%=1 TO M% : PRINT : NEXT I%
130 INPUT MENU$
135 IF MENU$="A" THEN GOSUB 4050
137 IF MENU$="B" THEN GOSUB 4000
139 IF MENU$="C" THEN GOSUB 4050
141 IF MENU$="D" THEN GOSUB 4000
143 IF MENU$="E" THEN GOSUB 4500
145 IF MENU$="F" THEN GOSUB 4500
170 PRINT "   Do you want to join the MENU again ? (Y or N)"
175 INPUT A$
180 IF A$="Y" THEN 75
935 END
940 REM- Subroutine 940-3650 to calculate all PO for given NN%
945 REM- Zero order product operator (unit matrix)
950 FOR M%=1 TO NS%
955 N%(0,M%)=M%
960 D%(O,M%)=1
965 MEXT M%
970 F%(0)=0
```

```
975  REM- First order product operators Ix, Iy, Iz
980  FOR K%=1 TO NN%
985  K1%=2^(K%-1) : K2%=K1%*K1%
990  F%(K2%)=0 : F%(2*K2%)=1 : F%(3*K2%)=0
995  M%=1 : SG%=1
1000 FOR C%=1 TO K1%
1010 N%(K2%,M%)=M%+SG%*K1%
1020 N%(2*K2%,M%)=M%+SG%*K1%
1030 N%(3*K2%,M%)=M%
1040 D%(K2%,M%)=1
1050 D%(2*K2,M%)=-SG%
1060 D%(3*K2%,M%)=SG%
1070 M%=M%+1
1080 NEXT C%
1090 SG%=-SG%
1100 IF M%<=NS% THEN 1000
1100 NEXT K%
2000 REM- Second order base operators (Two factor PO's)
2010 FOR KA%=1 TO NN%-1
2020 FOR KB%=KA%+1 TO NN%
2030 FOR JA%=1 TO 3
2040 FOR JB%=1 TO 4
2050 K2A%=4^(KA%-1)
2060 K2B%=4^(KB%-1)
2070 SA%=JA%*K2A% : SB%=JB%*K2B%
2080 GOSUB 5000
2090 NEXT JB%
2100 NEXT JA%
2110 NEXT KB%
2120 NEXT KA%
3000 REM- Three factor product operators
3010 IF NN%<3 THEN 3625
3020 FOR KC%=1 TO NN%-2
3030 FOR KD%=KC%+1 TO NN%-1
3040 FOR KB%=KD%+1 TO NN%
3050 FOR JC%=1 TO 3
3060 FOR JD%=1 TO 3
3070 FOR JB%=1 TO 3
3080 K2C%=4^(KC%-1) : K2D%=4^(KD%-1) : K2B%=4^(KB%-1)
3090 SC%=JC%*K2C% : SD%=JD%*K2D% : SB%=JB%*K2B%
3100 SA%=SC%+SD%
3100 GOSUB 5000
3120 NEXT JB%
3130 NEXT JD%
3140 NEXT JC%
3150 NEXTKB%
3160 NEXT KD%
```

```
3170 NEXT KC%
3500 REM- Four factor product operators
3510 IF NN%<4 THEN 3625
3520 FOR JC%=1 TO 3
3530 FOR JD%=1 TO 3
3540 FOR JE%=1 TO 3
3550 FOR JB%=1 TO
3560 SA%=JC%+4*JD%+16*JE%
3570 SB%=64*JB%
3580 GOSUB 5000
3590 NEXT JB%
3600 NEXT JE%
3610 NEXT JD%
3620 NEXT JC%
3625 REM- Operator labeling
3630 FOR S%=0 TO NS%^2-1
3635 GOSUB 3900
3637 NEXT S%
3640 REM- All product operators are now calculated and labeled
3641 REM- Arrays d(s,m,), n(s,m), F(s) and B$(s) are filled
3650 RETURN
3900 REM- Subroutine 3900-3045 generates label, given s
3901 B$="["
3904 R%=S%
3908 FOR I%=1 TO NN%
3912 J%(I%)=R%-4*INT(R%/4)
3916 R%=INT(R%/4)
3920 IF J%(I%)=0 THEN J$(I%)="1"
3924 IF J%(I%)=1 THEN J$(I%)="x"
3928 IF J%(I%)=2 THEN J$(I%)="y"
3932 IF J%(I%)=3 THEN J$(I%)="z"
3936 B$=B$+J$(I%)
3940 NEXT I%
3943 B$=B$+"]"
3945 RETURN
3950 REM- Subroutine 3950-3995 generates s, given label
3952 P=0 : S%=0
3955 FOR I%=1 TO NN%
3960 IF J$(I%)="1" THEN J%(I%)=0 GOTO 3985
3965 IF J$(I%)="X" THEN J%(I%)=1 GOTO 3985
3970 IF J$(I%)="Y" THEN J%(I%)=2 GOTO 3985
3975 IF J$(I%)="Z" THEN J%(I%)=3 GOTO 3985
3980 P=1 : GOTO 3995
3985 S%=S%+4^(I%-1)*J%(I%)
3990 NEXT I%
3995 RETURN
4000 REM- Subroutine 4000-4040 to output all product operators
```

```
4005 IF MENU$="D" THEN MENU$="C"
4010 FOR S%=0 TO NS%^2-1
4020 GOSUB 4155
4030 NEXT S%
4040 RETURN
4050 REM- Subroutine 4050-4110 to output one specified PO
4051 PRINT "Please label desired product operator, e.g. ";
4052 IF NN%=2 THEN B$="X,Y or 1,Z etc."
4053 IF NN%=3 THEN B$="X,Y,Z or 1,X,Y etc."
4054 IF NN%=4 THEN B$="X,Y,X,Z or X,Y,Z,1 etc."
4055 PRINT B$
4056 PRINT "  then press RETURN"
4060 IF NN%=2 THEN INPUT J$(1),J$(2)
4070 IF NN%=3 THEN INPUT J$(1),J$(2),J$(3)
4080 IF NN%=4 THEN INPUT J$(1),J$(2),J$(3),J$(4)
4090 GOSUB 3950
4095 IF P>0 THEN 4105
4100 GOSUB 4155 : GOTO 4107
4105 PRINT : PRINT : PRINT "Please try again "
4107 PRINT "Do you want another ? (Y or N)"
4108 INPUT A$ : IF A$="Y" THEN 4050
4110 RETURN
4155 REM- Subroutine 4155-4275 to display or print on PO, given s
4160 PRINT : PRINT B$(S%)+" =" : PRINT
4166 IF MENU$="C" THEN 4167 ELSE 4170
4167 LPRINT : LPRINT B$(S%)+" =" : LPRINT
4170 FOR M%=1 TO NS%
4180 FOR I%=1 TO N%(S%,M%)-1
4190 PRINT " .";
4195 IF MENU$="C" THEN 4196 ELSE 4200
4196 LPRINT " .";
4200 NEXT I%
4210 D%=D%(S%,M%)
4220 GOSUB 5300
4230 PRINT D$;
4235 IF MENU$="C" THEN 4236 ELSE 4240
4236 LPRINT D$;
4240 FOR I%=N%(S%,M%)=1 TO NS%
4250 PRINT " .";
4255 IF MENU$="C" THEN 4256 ELSE 4260
4256 LPRINT " .";
4260 NEXT I%
4265 PRINT
4266 IF MENU$="C" THEN LPRINT ELSE 4270
4270 NEXT M% : PRINT : PRINT
4272 IF MENU$="C" THEN 4273 ELSE 4275
4273 LPRINT : LPRINT
```

```
4275 RETURN
4400 PRINT "Do you want another ? (Y or N)" : INPUT H$
4410 IF H$="Y" THEN 4002
4500 REM- Subroutine 4500-4790 to express a given matrix
4501 REM- in terms of product operators
4530 FOR M%-0 TO NS% : FOR N%=0 TO NS%
4540 A(M%,N%)=0 : B(M%,N%)=0
4550 NEXT N% : NEXT M%
4555 PRINT "How many nonvanishing elements in the matrix ?"
4556 INPUT N
4560 PRINT "For every nonvanishing element in your matrix"
4561 PRINT "          d(m,n) = a + i*b"
4562 PRINT "please enter: m, n, a, b then press RETURN"
4569 FOR I%=1 TO N
4570 INPUT M%,N%,A,B
4580 A(M%,N%)=A : B(M%,N%)=B
4590 NEXT I%
4600 T$="Your matrix has the following non-zero elements"
4601 PRINT T$
4601 IF MENU$="F" THEN LPRINT T$
4610 FOR M%=1 TO NS% : FOR N%=1 TO NS%
4620 IF A(M%,N%)=0 THEN IF B(M%,N%)=0 THEN 4640
4625 B$="+i*" : B=B(M%,N%)
4626 IF B<0 THEN B=-B : B$="-i*"
4630 PRINT "d("M%","N%")=", A(M%,N%),B$;B
4635 IF NEMU$="F" THEN 4636 ELSE 4640
4636 LPRINT "d("M%","N%")=", A(M%,N%),B$;B
4640 NEXT N% : NEXT M%
4645 PRINT : PRINT : PRINT "Your matrix ="
4646 IF MENU$="F" THEN 4647 ELSE 4650
4647 LPRINT : LPRINT : LPRINT "Your matrix ="
4650 FOR S%=0 TO NS%^2-1
4660 R%=1-F%(S%)
4665 CR=0 : CI=0
4670 FOR M%=1 TO NS%
4680 N%=N%(S%,M%)
4690 CR=CR+(A(M%,N%)*R%+B(M%,N%%)*F%(S%))*D%(S%,M%)
4700 CI=CI+(B(M%,N%)*R%-A(M%,N%%)*F%(S%))*D%(S%,M%)
4710 NEXT M%
4720 CR=CR/2 : CI=CI/2
4730 IF CR=0 THEN IF CI=0 THEN 4780
4735 B$="+i" : B=CI
4736 IF CI<0 THEN B=-B : B$="-i*"
4740 PRINT "("CR,B$;B;")*"B$(S$)
4745 IF MENU$="F" THEN 4746 ELSE 4780
4746 LPRINT "("CR,B$;B;")*"B$(S$)
4780 NEXT S%
```

```
4781 PRINT : PRINT
4790 RETURN
5000 REM- Subroutine 5000-5090 special matrix multiplication
5010 S%=SA%+SB%
5020 FOR M%=1 TO NS%
5030 MB%=N%(SA%,M%)
5040 N%(S%,M%)=N%(SB%,MB%)
5050 D%(S%,M%)=D%(SA%,M%)*D%(SB%,MB%)5060 NEXT M%
5070 F%(S%)=F%(SA%)+F%(SB%)
5080 IF F%(S%)>1 THEN GOSUB 5100
5090 RETURN
5100 F%(S%)=0
5110 FOR M%=1 TO NS%
5120 D%(S%,M%)=-D%(S%,M%)
5122 NEXT M%
5130 RETURN
```

APPENDIX E: DEMONSTRATION OF THE ROTATION RULES

We demonstrate here the validity of the PO pulse rotations derived from the vector representation in Section II.6 (correspondence between the vector rotations and the PO formalism).

Demonstration of an αx rotation

We will demonstrate an α rotation around the x-axis:

$$I_y \xrightarrow{\alpha x} I_y \cos\alpha + I_z \sin\alpha \tag{E1}$$

Other rotations can be demonstrated in a similar way. We start from the rotation operator applied to a density matrix and we make use of the commutation rules for the operators I_x, I_y, I_z, which are:

$$\begin{aligned} I_x I_y - I_y I_x &= iI_z \\ I_y I_z - I_z I_y &= iI_x \\ I_z I_x - I_x I_z &= iI_y \end{aligned} \tag{E2}$$

It is necessary to emphasize that the above rules apply not only to an isolated spin but also to every particular nucleus in a multinuclear system (with or without coupling). In an AMX system for instance we have for nucleus A:

$$I_{xA} I_{yA} - I_{yA} I_{xA} = iI_{zA} \qquad \text{(and so on)}$$

Angular momentum components of *different* nuclei within a system are commutative. For instance:

$$\begin{aligned} I_{xA} I_{yM} - I_{yM} I_{xA} &= 0 \\ I_{xA} I_{xM} - I_{xM} I_{xA} &= 0 \end{aligned} \tag{E3}$$

The rotation operator $R_{\alpha x}$ [see(B45)] has the expression:

$$R_{\alpha x} = \exp(i\alpha I_x) \tag{E4}$$

Applied to a density matrix $D(n)$ it will yield:

$$D(n+1) = R_{\alpha x}^{-1} D(n) R_{\alpha x} \tag{E5}$$

In (E1) we have assumed that $D(n)$ is equal to I_y, so what we have to demonstrate is:

$$D(n+1) = R_{\alpha x}^{-1} I_y R_{\alpha x} = I_y \cos\alpha + I_x \sin\alpha \tag{E6}$$

In Appendix B we have demonstrated [see (B51)] that

$$R_{\alpha x} = \cos\frac{\alpha}{2}\cdot[\mathbf{1}] + i\sin\frac{\alpha}{2}\cdot(2I_x) \tag{E7}$$

Introducing this expression in (E6) gives

$$D(n+1) = \left[\cos\frac{\alpha}{2}\cdot[\mathbf{1}] - i\sin\frac{\alpha}{2}\cdot(2I_x)\right] I_y \left[\cos\frac{\alpha}{2}\cdot[\mathbf{1}] + i\sin\frac{\alpha}{2}\cdot(2I_x)\right]$$

$$= \cos^2\frac{\alpha}{2}\cdot I_y + 4\sin^2\frac{\alpha}{2}\cdot I_x I_y I_x + 2i\cos\frac{\alpha}{2} 4\sin\frac{\alpha}{2}(I_y I_x - I_x I_y)$$

After using the first relation in (E2) this becomes

$$D(n+1) = \cos^2\frac{\alpha}{2}\cdot I_y + 4\sin^2\frac{\alpha}{2}\cdot I_x I_y I_x + 2i\cos\frac{\alpha}{2} 4\sin\frac{\alpha}{2}(-iI_z)$$

$$= \cos^2\frac{\alpha}{2}\cdot I_y + 4\sin^2\frac{\alpha}{2}\cdot I_x I_y I_x + \sin\alpha\cdot I_z$$

If we use now the relation

$$I_x I_y I_x = -\frac{1}{4} I_y \tag{E8}$$

which will be demonstrated immediately, we get

$$D(n+1) = \cos\alpha\cdot I_y + \sin\alpha\cdot I_z$$

which confirms (E1).

To demonstrate (E8) we postmultiply the first relation in (E2) by I_x and, taking (B48) into account, we obtain

$$I_x I_y I_x - \frac{1}{4} I_y = iI_z I_x \tag{E9}$$

Premultiplying of the first relation in (E1) by I_x yields

$$\frac{1}{4} I_y - I_x I_y I_x = iI_x I_z \tag{E10}$$

Subtracting (E10) from (E9) gives

$$2I_xI_yI_x - \frac{1}{2}I_y = i(I_zI_x - I_xI_z) = i(iI_y) = -I_y$$

or

$$2I_xI_yI_x = -\frac{1}{2}I_y$$

which demonstrates (E8).

Rotation operators applied to product operators

Suppose we have to apply the rotation operator the $R_{\alpha xA}$ to the product operator [yzy]. The latter is a shorthand notation for the product.

$$(2I_{yA})(2I_{zM})(2I_{yX})$$

The subscripts A, M, X refer to the different nuclei in the system. Since these subscripts are omitted for simplicity in the product operator label [yzy], we have to keep in mind as a convention that the different nuclei of the system appear in the product operators always in the same order: A, M, X.

What we have to calculate is:

$$D(n+1) = \exp(-i\alpha I_{xA}) \cdot 8I_{yA}I_{zM}I_{yX} \exp(i\alpha I_{xA}) \qquad \text{(E11)}$$

We have stated (E3) that I_{xA} commutes with both I_{zM} and I_{yX}. This enables us to rewrite (E11) as:

$$D(n+1) = \exp(-i\alpha I_{xA}) \cdot I_{yA} \exp(i\alpha I_{xA}) 8I_{zM}I_{yX}$$

and we have reduced the problem to a known one. Using (E6) we get:

$$D(n+1) = 8(\cos\alpha \cdot I_{yA} + \sin\alpha \cdot I_{zA})I_{zM}I_{yX}$$

In shorthand notation:

$$[yzy] \xrightarrow{\alpha xA} [yzy]\cos\alpha + [zzy\}\sin\alpha \qquad \text{(E12)}$$

This can be phrased as follows: A rotation operator affects only one factor in the product operator and leaves the others unchanged. The affected factor parallels the vector rotation rules.

This is true in the case of a *selective pulse*. We have sometimes to handle *nonselective pulses*, affecting two or more nuclei in the system. In such cases, the procedure to follow is to substitute (in the calculations, not in the hardware) the nonselective pulse by a sequence of selective pulses following immediately one after another.

The problem

$$[yzy] \xrightarrow{\alpha xMX}$$

has to be handled as

$$[yzy] \xrightarrow{\alpha xM} [yzy]\cos\alpha - [yyy]\sin\alpha$$

$$\xrightarrow{\alpha xX} [yzy]\cos^2\alpha + [yzz]\cos\alpha\sin\alpha$$

$$-[yyy]\sin\alpha\cos\alpha - [yyz]\sin^2\alpha \qquad \text{(E13)}$$

The reader can easily check that the order in which αxM and αxX are applied is immaterial. The procedure described above has to be followed even if the spins affected by the pulse are magnetically equivalent.

The result (E13) may seem unexpectedly complicated for one single pulse. Fortunately, in most practical cases α is either 90° or 180°. In these cases, the procedure above leads to exhilariatingly simple results like:

$$[yzy] \xrightarrow{90xMX} -[yyz]$$

$$[yzy] \xrightarrow{180xMX} [yzy]$$

$$[xyz] \xrightarrow{90xAM} [xzz]$$

In these cases it is not necessary to split the non-selective pulse into subsequent selective pulses.

APPENDIX F: DEMONSTRATION OF THE COUPLED EVOLUTION RULES

Before going into the demonstration we need to point out two limitations:

a. It assumes I=1/2 for all nuclei in the system
b. It operates in the weak coupling case:
J (coupling constant)<< $\Delta\delta$ (chemical shift difference)

Between r.f. pulses the evolution of a two-spin system AX in the rotating frame is governed by the Hamiltonian.

$$\boldsymbol{H} = \hbar\left(\Omega_A I_{zA} + \Omega_X I_{zX} + 2\pi J I_{zA} I_{zX}\right) \qquad \text{(F1)}$$

The density matrix $D(n+1)$ at the end of the evolution is related to the initial matrix $D(n)$ as:

$$D(n+1) = \exp(-i\boldsymbol{H}t / \hbar)\cdot D(n)\cdot \exp(i\boldsymbol{H}t / \hbar) \qquad \text{(F2)}$$

Since all terms in (F1) commute with each other, we can write the evolution operator as:

$$\begin{aligned}\exp(i\boldsymbol{H}t / \hbar) &= \exp\left(i\Omega_A t I_{zA}\right)\exp\left(i\Omega_X t I_{zX}\right)\exp\left(i2\pi J t I_{zA} I_{zX}\right)\\ &= R_A R_X R_J\end{aligned}$$

where the order of the factors is immaterial. Relation (F2) can be rewritten as:

$$D(n+1) = R_J^{-1} R_X^{-1} R_A^{-1}\cdot D(n)\cdot R_A R_X R_J \qquad \text{(F3)}$$

In (F3) the actual coupled evolution is *formally* presented as the result of three independent evolutions due to shift A, shift X, and coupling J, respectively. In fact this is the way the coupled evolutions are handled in the PO formalism: as a succession of shift evolutions and evolutions due to spin-spin coupling (coupling evolutions).

The shift evolutions (noncoupled evolutions) are actually z-rotations and are easily handled with the vector rotation rules. Example:

$$\begin{aligned}[xy] &\xrightarrow{\Omega_A t} [xy]\cos\Omega_A t + [yy]\sin\Omega_A t\\ &\xrightarrow{\Omega_X t} [xy]\cos\Omega_A t\cos\Omega_X t - [xx]\cos\Omega_A t\sin\Omega_X t\\ &\qquad + [yy]\sin\Omega_A t\cos\Omega_X t - [yx]\sin\Omega_A t\sin\Omega_X t\end{aligned}$$

Or, with self-explanatory notations,

$$[xy] \xrightarrow{\Omega_A t} \xrightarrow{\Omega_X t} cc'[xy] - cs'[xx] + sc'[yy] - ss'[yx] \quad \text{(F4)}$$

The object of this appendix is to derive the rules for calculating

$$D(n+1) = R_J^{-1} D(n) R_J$$

where

$$R_J = \exp\left(i2\pi Jt I_{zA} I_{zX}\right) \quad \text{(F5)}$$

and $D(n)$ may be any of the product operators or a combination thereof.

We have to find first an expression similar to (B51) for the operator R_J. Calculating the powers of $I_{zA}I_{zX}$ we find

$$\left(I_{zA}I_{zX}\right)^n = \frac{1}{4^n}[\mathbf{1}] \qquad \text{for } n = \text{even} \quad \text{(F6)}$$

$$\left(I_{zA}I_{zX}\right)^n = \frac{1}{4^n} 4 I_{zA} I_{zX} \qquad \text{for } n = \text{odd} \quad \text{(F7)}$$

and this leads to

$$R_J = \cos\frac{\pi Jt}{2}[\mathbf{1}] + i\sin\frac{\pi Jt}{2} \cdot \left(4 I_{zA} I_{zX}\right) \quad \text{(F8)}$$

an expression we can use in calculating $D(n+1) = R_J^{-1} D(n) R_J$.

It is now the moment to introduce specific product operators for $D(n)$. We have to discuss three cases.

Case 1. Both nuclei A and X participate in the product operator with z or 1. Example:

$$D(n) = [zz] = \left(2 I_{zA}\right)\left(2 I_{zX}\right)$$

In this case $D(n)$ commutes with both I_{zA} and I_{zX} and this gives:

$$D(n+1) = R_J^{-1} D(n) R_J = D(n) R_J^{-1} R_J = D(n)$$

None of the POs [zz], [z1], [1z], [11] is affected by the coupling evolution. As a matter of fact, all these POs have only diagonal elements and are not affected by any evolution, shift or coupling.

Case 2. Both nuclei A and X participate in the product operator with an x or a y. Example:

$$D(n)=[xy]=(2I_{xA})(2I_{yX})$$

We can demonstrate that this kind of product operator also is not affected by the coupling evolution. In order to do so we have to take into account that, for I = 1/2, the components of the angular momentum ar anticommutative:

$$I_xI_y=-I_yI_x$$
$$I_yI_z=-I_zI_y \tag{F9}$$
$$I_zI_x=-I_xI_z$$

The validity of (F9) can be verified on the expressions (C1, C2) of the angular momentum components for I=1/2. Using (F8) to calculate $D(n+1)$ we have

$$D(n+1)=\left[\cos\frac{\pi Jt}{2}[\mathbf{1}]-i\sin\frac{\pi Jt}{2}\cdot(4I_{zA}I_{zX})\right]\times(4I_{xA}I_{yX})$$

$$\times\left[\cos\frac{\pi J}{2}[\mathbf{1}]+i\sin\frac{\pi J}{2}\cdot(4I_{zA}I_{zX})\right]$$

$$=\cos^2\frac{\pi Jt}{2}(4I_{xA}I_{yX})+\sin^2\frac{\pi Jt}{2}(4^3I_{zA}I_{zX}I_{xA}I_{yX}I_{zA}I_{zX})$$

$$+i\cos\frac{\pi Jt}{2}\sin\frac{\pi Jt}{2}(4^2)(I_{xA}I_{yX}I_{zA}I_{zX}-I_{zA}I_{zX}I_{xA}I_{yX})$$

Since the angular momentum components of A are commutative with those of X [see(E3)], we can rewrite the last result as

$$D(n+1)=\cos^2\frac{\pi Jt}{2}(4I_{xA}I_{yX})+\sin^2\frac{\pi Jt}{2}(4^3I_{zA}I_{xA}I_{zA}I_{zX}I_{yX}I_{zX})$$

$$+i\cos\frac{\pi Jt}{2}\sin\frac{\pi Jt}{2}(4^2)(I_{xA}I_{zA}I_{yX}I_{zX}-I_{zA}I_{xA}I_{zX}I_{yX}) \tag{F10}$$

Using (F9) we find out that the last parenthesis in (F10) is null. To calculate the product $I_{zA}I_{xA}I_{zA}I_{zX}I_{yX}I_{zX}$ in (F10) we use the following relations, similar to (E8) :

$$I_{zA}I_{xA}I_{zA} = -(I_{xA})/4 \quad ; \quad I_{zX}I_{yX}I_{zX} = -(I_{yX})/4$$

and we reduce (F10) to

$$D(n+1) = \cos^2\frac{\pi Jt}{2}\left(4I_{xA}I_{yX}\right) + \sin^2\frac{\pi Jt}{2}\left(4I_{xA}I_{yX}\right) = \left(4I_{xA}I_{yX}\right) = D(n)$$

All POs in the subset [*xx*],[*yx*],[*xy*],[*yy*] are not affected by the J evolution. Unlike the POs in Case 1, they *are* affected by the shift evolution (see F4). This is consistent with the fact that all the non-vanishing elements of these POs are double-quantum or zero-quantum coherences. The transition frequencies corresponding to these matrix elements do not contain the coupling J.

Case 3. The product operator exhibits z or 1 for one of the nuclei and x or y for the other nucleus. Only this kind of product operator is affected by the coupling. Example:

$$D(n) = [x1] = 2I_{xA}$$

Calculations similar to those performed in Case 2 lead to:

$$D(n+1) = \left[\cos\frac{\pi Jt}{2}[\mathbf{1}] - i\sin\frac{\pi Jt}{2}\cdot\left(4I_{zA}I_{zX}\right)\right] \times \left(2I_{xA}\right)$$

$$\times\left[\cos\frac{\pi Jt}{2}[\mathbf{1}] + i\sin\frac{\pi Jt}{2}\cdot\left(4I_{zA}I_{zX}\right)\right]$$

$$= \cos^2\frac{\pi Jt}{2}\left(2I_{xA}\right) + \sin^2\frac{\pi Jt}{2}\left(32I_{zA}I_{zX}I_{xA}I_{zA}I_{zX}\right)$$

$$+ i\cos\frac{\pi Jt}{2}\sin\frac{\pi Jt}{2}\left(8I_{xA}I_{zA}I_{zX} - 8I_{zA}I_{zX}I_{xA}\right)$$

$$= \cos^2\frac{\pi Jt}{2}\left(2I_{xA}\right) + \sin^2\frac{\pi Jt}{2}\left(32I_{zA}I_{xA}I_{zA}I_{zX}^2\right)$$

$$+ i\cos\frac{\pi Jt}{2}\sin\frac{\pi Jt}{2}\left(8I_{xA}I_{zA}I_{zX} - 8I_{zA}I_{xA}I_{zX}\right)$$

$$= \cos^2 \frac{\pi Jt}{2}(2I_{xA}) - \sin^2 \frac{\pi Jt}{2}(2I_{xA})$$

$$+ i \cos\frac{\pi Jt}{2} \sin\frac{\pi Jt}{2}(I_{xA}I_{zA} - I_{zA}I_{xA})(8I_{zX})$$

$$= \cos^2 \frac{\pi Jt}{2}(2I_{xA}) - \sin^2 \frac{\pi Jt}{2}(2I_{xA})$$

$$+ i \cos\frac{\pi Jt}{2} \sin\frac{\pi Jt}{2}(-iI_{yA})(8I_{zX})$$

$$= \cos \pi Jt(2I_{xA}) + \sin \pi Jt(4I_{yA}I_{zX})$$

$$= \cos \pi Jt[x1] + \sin \pi Jt[yz]$$

We have demonstrated that

$$[x1] \xrightarrow{J\ coupl} [x1]\cos \pi Jt + [yz]\sin \pi Jt$$

After doing similar calculations for all the POs in this category (i.e., [x1],[1x],[y1],[1y],[xz],[zx],[yz],[zy]), we can summarize the following rules for the evolution due to the coupling J_{AX}:

a. The coupling evolution operator R_J affects only those product operators in which one of the nuclei A,X is represented by x or y while the other is represented by 1 or z.

b. The effect of the J evolution is a rotation of x (or y) in the equatorial plane by πJt, while z is replaced by 1 and 1 by z in the new term. The format is:

PO after J evolution = cosπJt (former PO) + sinπJt (former PO in which x is replaced by y, y by $-x$, z by 1 and 1 by z). In systems with more than two nuclei, every nonvanishing coupling like J_{AM}, J_{AX}, J_{MX}, etc., has to be taken into account separately (the order is immaterial).

Note 1. From Appendices E and F it results that any rotation (r.f. pulse) or coupled evolution turns a given PO into a linear combination of POs within the basis set. In other words, if the density matrix can be expressed in terms of POs at the start of a sequence we will be able to express it as a combination of POs at any point of the sequence. This confirms that the PO basis set is a complete set.

Note 2. Moreover, in a coupled evolution, any x or y in the product operator can only become an x or y. Any z or 1 can only become a z or 1. This leads to a natural separation of the basis set (N^2 product operators) into N subsets of N operators each.

In the case of $N = 4$ (two nuclei) the four subsets are:

1) [11], [1*z*], [*z*1], [*zz*]

2) [*x*1], [*y*1], [*xz*], [*yz*]

3) [1*x*], [*zx*], [1*y*], [*zy*]

4) [*xx*], [*yx*], [*xy*], [*yy*]

In the case of N = 8 (three nuclei) the eight subsets are:

1) [111], [*z*11], [1*z*1], [*zz*1], [11*z*], [*z*1*z*], [1*zz*], [*zzz*]

2) [*x*11], [*y*11], [*xz*1], [*yz*1], [*x*1*z*], [*y*1*z*], [*xzz*], [*yzz*]

3) [1*x*1], [*zx*1], [1*y*1], [*zy*1], [1*xz*], [*zxz*], [1*yz*], [*zyz*]

4) [*xx*1], [*yx*1], [*xy*1], [*yy*1], [*xxz*], [*yxz*], [*xyz*], [*yyz*]

5) [11*x*], [*z*1*x*], [1*zx*], [*zzx*], [11*y*], [*z*1*y*], [1*zy*], [*zzy*]

6) [*x*1*x*], [*y*1*x*], [*xzx*], [*yzx*], [*x*1*y*], [*y*1*y*], [*xzy*], [*yzy*]

7) [*1xx*], [*zxx*], [1*yx*], [*zyx*], [1*xy*], [*zxy*], [1*yy*], [*zyy*]

8) [*xxx*], [*yxx*], [*xyx*], [*yyx*], [*xxy*], [*yxy*], [*xyy*], [*yyy*]

Under a coupled evolution, the descendents of a PO are to be found only within its own subset.

APPENDIX G: PO EVOLUTION TABLES

The tables below summarize the effect of a coupled evolution on each of the sixteen POs of an AX system (two spin 1/2 nuclei). The first column in each table indicates the PO before the evolution, while the next columns indicate the newly created POs (including the initial one). A coupled evolution implies three PO operations: shift A, shift X, coupling *J*. Nevertheles no more than four terms are generated from the initial one: when both shifts are active the coupling is not.

a. First subset (not affected by evolution)

Initial PO	Final POs			
[11]	[11]	0	0	0
[z1]	0	[z1]	0	0
[1z]	0	0	[1z]	0
[zz]	0	0	0	[zz]

b. Second subset (affected by A shift and *J* coupling)

Initial PO	Final POs			
[x1]	$cC[x1]$	$sC[y1]$	$-sS[xz]$	$cS[yz]$
[y1]	$-sC[x1]$	$cC[y1]$	$-cS[xz]$	$-sS[yz]$
[xz]	$-sS[x1]$	$cS[y1]$	$cC[xz]$	$sC[yz]$
[yz]	$-cS[x1]$	$-sS[y1]$	$-sC[xz]$	$cC[yz]$

c. Third subset (affected by X shift and J coupling)

Initial PO	Final POs			
$[1x]$	$c'C[1x]$	$-s'S[zx]$	$s'C[1y]$	$c'S[zy]$
$[zx]$	$-s'S[1x]$	$c'C[zx]$	$c'S[1y]$	$s'C[zy]$
$[1y]$	$-s'C[1x]$	$-c'S[zx]$	$c'C[1y]$	$-s'S[zy]$
$[zy]$	$-c'S[1x]$	$-s'C[zx]$	$-s'S[1y]$	$c'C[zy]$

d. Fourth subset (affected by A shift and X shift)

Initial PO	Final POs			
$[xx]$	$cc'[xx]$	$sc'[yx]$	$cs'[xy]$	$ss'[yy]$
$[yx]$	$-sc'[xx]$	$cc'[yx]$	$-ss'[xy]$	$cs'[yy]$
$[xy]$	$-cs'[xx]$	$-ss'[yx]$	$cc'[xy]$	$sc'[yy]$
$[yy]$	$ss'[xx]$	$-cs'[yx]$	$-sc'[xy]$	$cc'[yy]$

$c = \cos\Omega_A t$; $c' = \cos\Omega_X t$; $C = \cos\pi Jt$

$s = \sin\Omega_A t$; $s' = \sin\Omega_X t$; $S = \sin\pi Jt$

APPENDIX H: DEMONSTRATION OF THE REFOCUSING RULES

We demonstrate here the validity of the rules stated in Section II.8 to handle a refocusing routine:

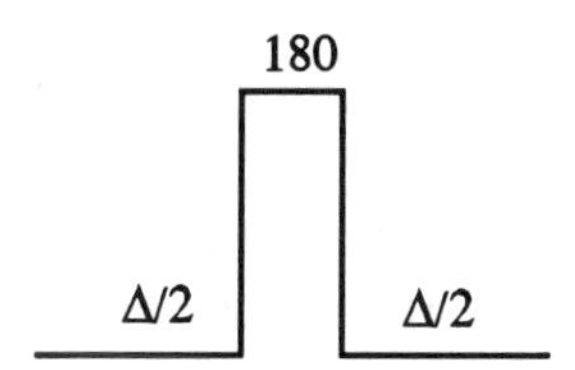

When handling this segment of a sequence in the conventional way we have to subject the density matrix $D(n)$ to a string of operators representing the $\Delta/2$ evolution (shifts and couplings), the r.f. pulse (R_{180}), then again the $\Delta/2$ evolution. For two nuclei the string would be:

$$R = R_A R_X R_{AX} R_{180} R_{AX} R_X R_A \tag{H1}$$

where R_A, R_X are shift operators and R_{AX} is the coupling operator.

All shift and coupling evolution operators commute with each other. In order to simplify the expression (H1) we have to find out how they commute with R_{180}. Let us concentrate on one nucleus (e.g., nucleus A) and see how R_A commutes with R_{180xA}. According to relation (B51)

$$R_{180xA} = \exp(i\pi I_{xA}) = \cos\frac{\pi}{2}[\mathbf{1}] + i\sin\frac{\pi}{2}(2I_{xA}) = 2iI_{xA} \tag{H2}$$

A similar expression can be written for the shift evolution operator R_A which represents a rotation by $\alpha = \Omega_A \Delta/2$ around the z axis.

$$R_A = R_{\alpha A} = \exp(i\alpha I_{zA}) = \cos\frac{\alpha}{2}[\mathbf{1}] + i\sin\frac{\alpha}{2}(2I_{zA}) \tag{H3}$$

We want to examine the product of the two operators:

$$R_{180xA}R_{\alpha A} = 2iI_{xA}\left[\cos\frac{\alpha}{2}[\mathbf{1}] + i\sin\frac{\alpha}{2}(2I_{zA})\right]$$

Using the anticommutativity of I_x and I_z for I=1/2 stated in (F10) we can rewrite the product as

$$R_{180xA}R_{\alpha A} = \left[\cos\frac{\alpha}{2}[\mathbf{1}] - i\sin\frac{\alpha}{2}(2I_{zA})\right]2iI_{xA} = R_{(-\alpha)A}R_{180xA}$$

We notice that $R_{(-\alpha)A}$ is the reciprocal of $R_{\alpha A}$ since their combined action would leave the density matrix unaffected. We can therefore write

$$R_{180xA}R_{\alpha A} = R_{\alpha A}^{-1}R_{180xA} \tag{H4}$$

and this is the commutation rule we needed.

One can check that the rule is the same if instead of R_{180xA} we use R_{180yA} or $R_{180\Phi A}$ (180° rotation about an arbitrary axis in the *xy* plane). For the last case one has use

$$R_{180\Phi A} = R_{(-\Phi)zA}R_{180xA}R_{\Phi zA} = R_{180\Phi A}R_{\Phi zA}^{2} = R_{180xA}R_{(2\Phi)zA}$$

$$= 2iI_{xA}\left(\cos\Phi\cdot[\mathbf{1}] + i\sin\Phi\cdot 2I_{zA}\right) = \cos\Phi\cdot 2iI_{xA} + i\sin\Phi\cdot(-2iI_{yA})$$

$$= R_{180xA}\cos\Phi + R_{180yA}\sin\Phi$$

We have used here $I_{xA}I_{zA} = -iI_{yA}/2$ which is a consequence of (E2) and (F9).

Therefore we can write in general

$$R_{180A}R_{A} = R_{A}^{-1}R_{180xA} \tag{H5}$$

On the other hand we have

$$R_{180A}R_{X} = R_{X}R_{180xA} \tag{H6}$$

since operators acting on different nuclei always commute.

A similar pattern can be followed to demonstrate

$$R_{180A}R_{AX} = R_{AX}^{-1}R_{180A} \tag{H7}$$

$$R_{180X}R_{AX} = R_{AX}^{-1}R_{180X} \tag{H8}$$

With these commutation rules we now can rearrange a string like the one in (H1):

$$R = R_A R_X R_{AX} R_{180\,A} R_{AX} R_X R_A = R_A R_X R_{AX} R_{AX}^{-1} R_X R_A^{-1} R_{180\,A}$$
$$= R_X^2 R_{180A} = R_{180A} R_X^2$$

Only the shift X is expressed in the final result, while shift A and coupling AX are refocused. Since R_X is the shift evolution for Δ/2,

$$R_X^2$$

is the operator for the full delay Δ.

If the 180° pulse is applied on both nuclei A and X, the rearranging yields:

$$R = R_A R_X R_{AX} R_{180A} R_{180X} R_{AX} R_X R_A = R_A R_X R_{AX} R_{180A} R_{AX}^{-1} R_X^{-1} R_A R_{180X}$$
$$= R_A R_X R_{AX} R_{AX} R_X^{-1} R_A^{-1} R_{180A} R_{180X} = R_{AX}^2 R_{180A} R_{180X} = R_{180A} R_{180X} R_{AX}^2$$

Both shifts are refocused, the coupling only is expressed in the final result. This confirms the refocusing rules stated in Section II.8. There is no difficulty in extending them to more than two nuclei.

APPENDIX I: SUPPLEMENTARY DISCUSSIONS

2DHETCOR without Δ_1

It is stated in Part I, Section 3.7, that in order to understand the role of the delay Δ_1 (see Figure I.2) we should carry on the calculations without it. We do this here.

We start by writing the density matrix before the combined $90xCH$ pulse:

$$D(5)=\begin{bmatrix} 3 & 0 & F & 0 \\ 0 & 2 & 0 & G \\ F* & 0 & 3 & 0 \\ 0 & G* & 0 & 2 \end{bmatrix} \tag{I1}$$

If we have the delay Δ_1 = 1/2J, the expressions for F and G are taken from (I.32):

$$\begin{aligned} F &= d_{13}(5) = -2\exp[-i\Omega_H(t_e+\Delta_1)] \\ G &= d_{24}(5) = -F \end{aligned} \tag{I2}$$

If Δ_1 is not used, the expressions for F and G, taken from (I.25) and (I.26), are:

$$F = d_{13}(4) = -2i\exp(-i\Omega_H t_e) = G \tag{I3}$$

We apply the operator R_{90xCH} given in (I.34) to D(5) in (I1).

$$D(5)R_{90xCH} = \frac{1}{2}\begin{bmatrix} 3+iF & 3i-F & 3i+F & -3+iF \\ 2i-G & 2+iG & -2+iG & 2i+G \\ 3i+F* & -3+iF* & 3+iF* & 3i-F* \\ -2+iG* & 2i+G* & 2i-G* & 2+iG* \end{bmatrix}$$

Premultiplication with R_{90xCH}^{-1} gives the following expression for D(7)

$$\frac{1}{4}\begin{bmatrix}
\begin{array}{l}10+i(F+G\\ -F^*-G^*)\end{array} & \begin{array}{l}2i-F+G\\ +F^*-G^*\end{array} & \begin{array}{l}F+G\\ +F^*+G^*\end{array} & \begin{array}{l}i(F-G\\ +F^*-G^*)\end{array} \\
\begin{array}{l}-2i+F-G\\ -F^*+G^*\end{array} & \begin{array}{l}10+i(F+G\\ -F^*-G^*)\end{array} & \begin{array}{l}-i(F-G\\ +F^*-G^*)\end{array} & \begin{array}{l}F+G\\ +F^*+G^*\end{array} \\
\begin{array}{l}F+G\\ +F^*+G^*\end{array} & \begin{array}{l}i(F-G\\ +F^*-G^*)\end{array} & \begin{array}{l}10-i(F+G\\ -F^*-G^*)\end{array} & \begin{array}{l}2i+F-G\\ -F^*+G^*\end{array} \\
\begin{array}{l}-i(F-G\\ +F^*-G^*)\end{array} & \begin{array}{l}F+G\\ +F^*+G^*\end{array} & \begin{array}{l}-2i-F+G\\ +F^*-G^*\end{array} & \begin{array}{l}10-i(F+G\\ -F^*-G^*)\end{array}
\end{bmatrix}$$

(I4)

Now we can see why having $G = F$ is counterproductive. The factors F and G contain the proton information [see (I3)] and, when they are equal, this information disappears from the observable single-quantum coherences d_{12} and d_{34}. In this case $D(7)$ becomes

$$D(7)=\frac{1}{2}\begin{bmatrix}
5+iF-iF^* & 2i & F+F^* & 0 \\
-2i & 5+iF-iF^* & 0 & F+F^* \\
F+F^* & 0 & 5-iF-iF^* & 2i \\
0 & F+F^* & -2i & 5-iF-iF^*
\end{bmatrix}$$

(I5)

Should be i's not 2i's!

No pulse follows after $t(7)$ and each matrix element evolves in its own "slot." The matrix elements d_{12} and d_{34}. will be affected by the carbon evolution during the detection t_d (preceded or not by the coupled evolution Δ_2) but they will not be proton modulated. We will not have a 2D.

The DM calculations in Sections 3.6 to 3.9 have been carried on *with* the delay Δ_1 = 1/2J, i.e, with G = – F [see (I2)]. In this case D(7) is

$$D(7) = \frac{1}{2}\begin{bmatrix} 5 & i - F + F^* & 0 & i(F + F^*) \\ -i + F - F^* & 5 & -i(F + F^*) & 0 \\ 0 & i(F + F^*) & 5 & i + F - F^* \\ -i(F + F^*) & F + F^* & -i - F + F^* & 5 \end{bmatrix} \tag{I6}$$

Taking F from (I2) and using the notations (I.31)

$$c = \cos[\Omega_H(t_e + \Delta_1)]$$
$$s = \sin[\Omega_H(t_e + \Delta_1)]$$

we obtain

$$F = -2(c - is)$$
$$F - F^* = 4is \tag{I7}$$
$$F + F^* = -4c$$

By introducing (I7) into (I6) we can verify the expression of $D(7)$ given in (I.35). The carbon observables d_{12} and d_{34} contain now the factor s which carries the proton information. The role of the subsequent coupled evolution Δ_2 is explained in Section 3.9.

2DHETCOR without the 180xC pulse

We demonstrate here the result (I.51). The density matrix at $t(1)$, taken from (I.12), is

$$D(1)=\begin{bmatrix} 2 & 0 & -2i & 0 \\ 0 & 3 & 0 & -2i \\ 2i & 0 & 2 & 0 \\ 0 & 2i & 0 & 3 \end{bmatrix} \tag{I8}$$

Since the 180xC pulse and the delay Δ_1 are suppressed, all we have between $t(1)$ and $t(5)$ is an evolution t_e. The density matrix just before the 90xCH pulse is then

$$D(5)=\begin{bmatrix} 2 & 0 & B & 0 \\ 0 & 3 & 0 & C \\ B^* & 0 & 2 & 0 \\ 0 & C^* & 0 & 3 \end{bmatrix} \tag{I9}$$

with

$$\begin{aligned} B &= -2i\exp(-i\Omega_{13}t_e) \\ C &= -2i\exp(-i\Omega_{24}t_e) \end{aligned} \tag{I10}$$

We apply the 180xCH operator to (I9).

$$D(5)R_{90xCH}=\frac{1}{2}\begin{bmatrix} 2+iB & 2i-B & 2i+B & -2+iB \\ 3i-C & 3+iC & -3+iC & 3i+C \\ 2i+B^* & -2+iB^* & 2+iB^* & 2i-B^* \\ -3+iC^* & 3i+C^* & 3i-C^* & 3+iC^* \end{bmatrix}$$

Premultiplication with R^{-1}_{90xCH} gives the following expression for D(7)

$$\frac{1}{4}\begin{bmatrix} 10+i(B+C-B^*-C^*) & 2i-B+C+B^*-C^* & B+C+B^*+C^* & i(B-C+B^*-C^*) \\ -2i+B-C-B^*+C^* & 10+i(B+C-F^*-C^*) & -i(B-C+B^*-C^*) & B+C+B^*+C^* \\ B+C+B^*+C^* & i(B-C+B^*-C^*) & 10-i(B+C-B^*-C^*) & 2i+B-C-B^*+C^* \\ -i(B-C+B^*-C^*) & B+C+B^*+C^* & -2i-B+C+B^*-C^* & 10-i(B+C-B^*-C^*) \end{bmatrix}$$

(I11)

We follow now the evolution of the carbon single-quantum coherences.

$$d_{12}(7)=\frac{1}{4}(-2i-B+C+B^*-C^*)$$

$$d_{34}(7)=\frac{1}{4}(-2i+B-C-B^*+C^*)$$

$$d_{12}(7)+d_{34}(7)=-i \tag{I12}$$

The terms B and C, which contain the proton information, are absent from the sum. If we start the acquisition at $t(7)$, with the decoupler on, we will not have a 2D. This is why Δ_2 still is necessary.

Proceeding as in (I.36) to (I.43) we have:

$$\begin{aligned} d_{12}(8) &= d_{12}(7)\exp(-i\Omega_C\Delta_2)\exp(-i\pi J\Delta_2) \\ &= -id_{12}(7)\exp(-i\Omega_C\Delta_2) \\ d_{34}(8) &= +id_{34}(7)\exp(-i\Omega_C\Delta_2) \end{aligned} \tag{I13}$$

Treating the detection as in (I.44) to (I.47) we obtain

$$d_{12}(9) = -id_{12}(7)\exp(-i\Omega_C\Delta_2)\exp(-i\Omega_C t_d)$$
$$d_{34}(9) = +id_{12}(7)\exp(-i\Omega_C\Delta_2)\exp(-i\Omega_C t_d) \qquad \text{(I14)}$$

$$M_{TC}(9) = M_{oC}\left[d_{12}^*(9) + d_{34}^*(9)\right]$$
$$= -iM_{oC}\left[d_{12}^*(7) - d_{34}^*(7)\right]\exp(i\Omega_C\Delta_2)\exp(i\Omega_C t_d)$$
$$= -i(M_{oC}\ /\ 2)(B - B^* - C + C^*)\exp\left[i\Omega_C(t_d + \Delta_2)\right] \qquad \text{(I15)}$$

From (I10) we have

$$B - B^* = -2i(\cos\Omega_{13}t_e - i\sin\Omega_{13}t_e) - 2i(\cos\Omega_{13}t_e + i\sin\Omega_{13}t_e)$$
$$= -4i\cos\Omega_{13}t_e \qquad \text{(I16)}$$
$$C - C^* = -4i\cos\Omega_{24}t_e \qquad \text{(I17)}$$
$$M_{TC}(9) = -i(M_{oC}\ /\ 2)(-4i\cos\Omega_{13}t_e + 4i\cos\Omega_{24}t_e)\exp\left[i\Omega_C(t_d + \Delta_2)\right]$$
$$= -2M_{oC}(\cos\Omega_{13}t_e - \cos\Omega_{24}t_e)\exp\left[i\Omega_C(t_d + \Delta_2)\right]$$

which confirms (I.51).

Fully coupled 2DHETCOR

If the proton decoupler is not turned on during the detection, the matrix elements d_{12} and d_{34} ,will evolve with different frequencies in the domain t_d. Each of them is proton modulated, even if their sum at time $t(7)$ is not, and this renders the delay Δ_2 unnecessary. The detection starts at $t(7)$ and we will have

$$M_{TC}(9) = -M_{oC}\left[d_{12}^*(7)\exp(i\Omega_{12}t_d) + d_{34}^*(7)\exp(i\Omega_{34}t_d)\right]$$

By introducing (I12), (I16), and (I17) in the expression above we obtain

$$M_{TC}(9) = -iM_{oC}(1/2 - \cos\Omega_{13}t_e + \cos\Omega_{24}t_e)\exp(i\Omega_{12}t_d)$$
$$- iM_{oC}(1/2 + \cos\Omega_{13}t_e - \cos\Omega_{24}t_e)\exp(i\Omega_{34}t_d)$$

which confirms (I.52).

APPENDIX J: PRODUCT OPERATORS AND MAGNETIZATION COMPONENTS

Finding the magnetization components

Suppose that at a given moment $t(n)$ of the sequence we have the density matrix $D(n)$ expressed in terms of product operators. The x magnetization components for nuclei A, M, X (we take a three-spin system as an example) are given by:

$$M_{xA}(n) = -(M_{oA} / p'_A)(\text{ coefficient of } [x11] \text{ in } D(n))$$
$$M_{xM}(n) = -(M_{oM} / p'_M)(\quad '' \quad [1x1] \quad '' \quad D(n)) \qquad \text{(J1)}$$
$$M_{xX}(n) = -(M_{oX} / p'_X)(\quad '' \quad [11x] \quad '' \quad D(n))$$

with similar expressions for M_y and M_z.

M_{oA}, M_{oM}, M_{oX} are the equilibrium magnetizations for the respective nuclei. The factor p' is related to the Boltzmann factor and has the expression

$$p'_A = p_A / 2N = \hbar\Omega_A / 2NkT \qquad \text{(J2)}$$

with similar expressions for p'_M, p'_X (for two nuclei we adopt p, q instead of p_A, p_X). N = total number of quantum states in the system (degenerate or not). For m nuclei with $I = 1/2$ we have $N = 2^m$.

The quantity $1/N$ represents the average population per state as the total population is normalized to 1. The above procedure is justified by the relation (B17), keeping in mind the orthogonality of product operators (relation D1).

Fast and slow magnetization components

In the first part of this book, when deriving the magnetization components from the density matrix elements, we found it instructive to calculate separately the *fast* and *slow* components of the transverse magnetization (see Section I.3.9). This can be achieved in the PO formalism as well.

For an AX system (two spin 1/2 nuclei), one can split the M_{xA} magnetization into M_{x12} (fast) and M_{x34} (slow) as follows:

$$M_{x12} = -\frac{M_{oA}}{2p'}(\text{coeff. of } [x1] + \text{coeff. of } [xz])$$

$$M_{x34} = -\frac{M_{oA}}{2p'}(\text{coeff. of } [x1] - \text{coeff. of } [xz])$$

with similar relations for M_y.

For nucleus X the fast and slow components are:

$$M_{x13} = -\frac{M_{oA}}{2p'}(\text{coeff. of } [1x] + \text{coeff. of } [zx])$$

$$M_{x24} = -\frac{M_{oA}}{2p'}(\text{coeff. of } [1x] - \text{coeff. of } [zx])$$

Using Table II.1, one can check that

$$([1x]+[zx])/2 \quad \text{and} \quad ([1x]-[zx])/2$$

represent the matrix elements d_{13} and d_{24}, respectively.

Writing the initial density matrix

We assume usually that, before the pulse sequence starts, all spins are at thermal equilibrium with the lattice.

With the sign convention adopted in this book (vide infra) the density matrix $D(0)$ will be:

$$D(0) = -p'_A[z11] - p'_M[1z1] - p'_X[11z] \qquad \text{(J3)}$$

This is consistent with

$$M_{zA} = M_{oA} \quad ; \quad M_{xA} = 0 \quad ; \quad M_{yA} = 0$$

and is the same for spins M and X.

If we work in a steady state in which one of the nuclei (let's say nucleus A) does not have enough time to recover from the previous cycle, we can start with

$$D(0) = -\lambda p'_A[z11] - p'_M[1z1] - p'_X[11z] \qquad \text{(J4)}$$

where $\lambda \leq 1$. If M_{zA} has been brought to zero by the last pulse of the sequence and the start of the next sequence comes after a delay d, then

$$\lambda = 1 - \exp(-d / T_{1A})$$

Generally we start a sequence with no transverse magnetization; this implies that d is long enough with respect to T_2 for all nuclei.

Sign convention

Throughout this book the conventional positive sense (right hand rule) has been adopted for all rotations (due to pulses or evolutions). For instance a rotation about Oz brings the magnetization from x to y, to $-x$, to $-y$. A rotation about Ox goes:

$$y \text{ to } z \text{ to } -y \text{ to } -z$$

This is the actual sense of precession when the magnetogyric ratio, γ, is negative. It has been assumed that the main magnetic field B_0 is directed upwards (along $+Oz$) and thus at equilibrium the magnetization is directed up and the angular momentum is directed downwards (along $-Oz$) which is also consistent with a negative γ.

This way, the most populated state is the one with the angular momentum negative. Most features of the NMR sequences (one- and two-dimensional) do not depend on the sign of γ for all practical purposes. Commercial NMR spectrometers do not provide specific means to distinguish the sign of γ. Therefore the above rules can be applied for all nuclei in the system, no matter what the sign of their magnetogyric ratio.

If for special purposes the sign of γ has to be taken into account, then for positive γ we have to use left hand rotation rules and start at thermal equilibrium with I_z directed upwards.

APPENDIX K: WHEN TO DROP NONOBSERVABLE TERMS (NOT)

In the expression of the density matrix we are interested in those POs which represent observable magnetization components: M_x and M_y for the nucleus (or nuclei) which are observed (see Appendix J). We have to carry all the nonobservable terms through the calculations as long as there is a possibility for them to generate observable terms (following a pulse or an evolution).

It is useful to be aware when it is safe to drop the nonobservable terms, or just include them in the nondescript designation NOT.

Rule #1. In the final expression $D(n)$ of the density matrix we have to write down explicitly the observable terms only for the specific nucleus which is observed. This includes the POs which show x or y for the observed nucleus and 1 for all others.

Rule #2. A decoupled evolution does not generate observable POs out of NOT or the reverse: it merely replaces x by y or y by $-x$. So, if the last event of the sequence (the detection) is a decoupled evolution, we can do the selection earlier, when writing $D(n-1)$.

Rule #3. A coupled evolution can interchange x and y but also 1 and z. Although, we can do some term dropping before the last evolution, even if coupled. In writing $D(n-1)$ we will retain only the POs which contain x or y for the nucleus to be observed and z or 1 for the other nuclei. Everything else is a NOT.

For instance, if we observe nucleus A (in an AMX system), the following terms must be kept:

$$[x11], [x1z], [xz1], [xzz], [y11], [y1z], [yz1], [yzz].$$

Note. If a pulse still follows, it is recommended to use utmost care in dropping terms. An experienced student will find for example that if the observable is A and the last pulse is on X, terms like $[1x]$, $[1z]$, $[zx]$, (no x or y in position A) can be labeled as NOT before the pulse since an X pulse will never render them observable.

APPENDIX L: MAGNETIC EQUIVALENCE, THE MULTIPLET FORMALISM

While the PO formalism is basically a shorthand notation for density matrix calculations, we introduce here an even more compact notation (we call it the the multiplet formalism) to simplify the PO treatment of CH_n systems. In such a system n nuclei out of n+1 are magnetically equivalent.

The density matrix at thermal equilibrium for a CH_3 system is:

$$D(0) = -p'[z111] - q'([1z11] + [11z1] + [111z]) \qquad \text{(L1)}$$

We introduce the notation $\{1z\}$ for the sum in parentheses and write $D(0)$ as:

$$D(0) = -p'[z111] - q'\{1z\} \qquad \text{(L2)}$$

The notation $\{1z\}$ is not only shorter but it allows us to write the last sum without specifying the number of protons in the system CH_n. We make the convention that:

$$\{1z\} = [1z11] + [11z1] + [111z] \qquad \text{for } CH_3$$

$$\{1z\} = [1z1] + [11z] \qquad \text{for } CH_2$$

$$\{1z\} = [1z] \qquad \text{for } CH$$

Other examples of the { } notation:

$$\{1x\} = [1x1] + [11x] \qquad \text{for } CH_2$$

$$\{zy\} = [zy11] + [z1y1] + [z11y] \qquad \text{for } CH_3$$

The sum { } has always n terms where n is the number of magnetically equivalent nuclei. In order to be consistent we have to write $\{z1\}$ as:

$$\{z1\} = [z111] + [z111] + [z111] = 3[z111] \quad \text{for } CH_3$$

$$\{z1\} = 2[z11] \qquad \text{for } CH_2$$

$$\{z1\} = [z1] \qquad \text{for } CH$$

With this convention we can rewrite (L2) as:

$$D(0) = -(p'/n)\{z1\} - q'\{1z\} \qquad \text{(L3)}$$

an expression valid for any *n*.

Let us apply a 90*xX* pulse to *D*(0).

$$D(0) \xrightarrow{90xX} -(p'/n)\{z1\} + q'\{1y\} \qquad \text{(L4)}$$

The expression above is easily obtained through the conventional PO rules applied to the expression (L1) or its equivalents for CH_2 or CH.

The above example may inspire us to hope that any sequence involving a CH_2 or CH_3 system may be treated as a CH, just by replacing [ ] with { }. Unfortunately this is not always true. It is important for the user of this formalism to be aware of how far the { } notations can be used and when we have to give up and go back to the explicit [ ] notation.

Rule #1

Not all the POs in the basis set can be included in { } sums. For instance [1*xz*] is not a legitimate term of such a sum. No more than one of the protons may come in the PO with *x*, *y* or *z*, the others must participate with a 1.

Rule #2

Rotations due to r.f. pulses do not break the { } formalism and can be treated according to the vectorial model. We have done this in (L4).

Rule #3

Likewise, *noncoupled* evolutions (which are merely *z*-rotations) can be treated according to the vectorial representation. Example:

$$\{zx\} \xrightarrow{shift\ X} \{zx\}\cos\Omega_X t + \{zy\}\sin\Omega_X t$$

We can satisfy ourselves that this is true in the case of CH_2:

$$\{zx\} = [zx1] + [z1x]$$

$$\xrightarrow{shift\ X} [zx1]\cos\Omega_X t + [zy1]\sin\Omega_X t + [z1x]\cos\Omega_X t + [z1y]\sin\Omega_X$$

$$= \{zx\}\cos\Omega_X t + \{zy\}\sin\Omega_X t$$

Rule #4

A coupled evolution usually breaks the { } formalism, since it generates PO's that cannot be included in { } sums. A fortunate exception we can take advantage of is presented by the sums: $\{1x\}$, $\{1y\}$, $\{zx\}$, $\{zy\}$. The coupled evolution of any of these sums does not break the { } formalism and can be treated according to Section II.7.

Rule #5

If the density matrix contains sums as $\{x1\}$, $\{yz\}$ or $\{xy\}$ and a coupled evolution follows, we have to give up the { } shorthand and go back to the regular PO formalism ([ ] notation).

Even here there is an exception and it is worth talking about because it may save a lot of term writing. The exception works when the coupled evolution we are talking about is the last act of the sequence or is just followed by a decoupled evolution. In this case we are interested in the observable terms only (see Appendix K) and the coupled evolution yields unexpectedly simple results (we observe nucleus A):

$$\{x1\} \longrightarrow cC^n\{x1\} + sC^n\{y1\} + \text{NOT}$$

$$\{y1\} \longrightarrow -sC^n\{x1\} + cC^n\{y1\} + \text{NOT}$$

$$\{xz\} \longrightarrow -sSC^{n-1}\{x1\} + cSC^{n-1}\{y1\} + \text{NOT}$$

$$\{yz\} \longrightarrow -cSC^{n-1}\{x1\} - cSC^{n-1}\{y1\} + \text{NOT}$$

$$\{xx\} \longrightarrow \text{NOT}$$

$$\{yx\} \longrightarrow \text{NOT}$$

$$\{xy\} \longrightarrow \text{NOT}$$

$$\{yy\} \longrightarrow \text{NOT}$$

$c = \cos \Omega_C t$; $C = \cos \pi J t$; n = number of protons

$s = \sin \Omega_C t$; $S = \sin \pi J t$; NOT = nonobservable terms

The above evolution rules can be verified by writing the product operators in the conventional way and calculating the evolution of the CH_n system for n=1, 2, and 3, separately.

Rule #6

The procedure for retrieving the magnetization components in the { } formalism differs only slightly from that stated in Appendix J.

$$M_{xC} = -(nM_{oC} / p') \times (\text{coefficient of } \{x1\})$$
$$M_{xH} = -(M_{oH} / q') \times (\text{coefficient of } \{1x\}) \qquad \text{(L5)}$$

with similar expressions for M_y and M_z. M_{oH} is the equilibrium magnetization due to all n magnetically equivalent protons.

APPENDIX M: ROTATIONS ABOUT NONTRIVIAL AXES

Rotations due to r.f. pulses follow in the PO formalism the pattern of vector kinematics. In Section II.6 we have discussed rotations about the axes *Ox* and *Oy*. In order to treat the general case (rotation axis with any spatial orientation) we use spherical coordinates as shown in Figure M.1. The rotation axis makes an angle θ with *Oz* ($0 \leq \theta \leq 180^\circ$) and its horizontal projection makes the angle Φ with *Ox* (positive sense for Φ is from *Ox* toward *Oy*).

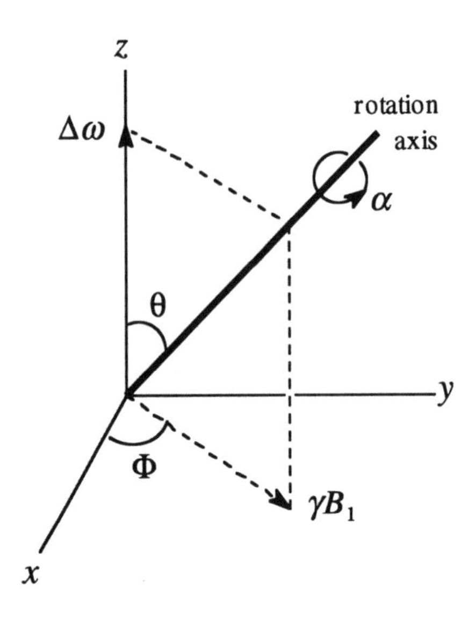

Figure M.1. Rotation axis in spherical coordinates.

If the r.f. pulse is applied on resonance, $\theta = 90°$. For a deviation $\Delta\omega = \gamma B_o - \omega_{tr}$ from resonance, the angle θ is different from 90°; its value is defined by $\tan\theta = \gamma B_1/\Delta\omega$. The angle Φ is the transmitter phase which is always taken with respect to an initially established receiver phase.

The result of a rotation by an angle α (right hand rule) about the axis defined above is described in Table M.1.

Table M.1. Final Angular Momentum Components after Rotation by an Angle α about the Rotation Axis Defined by Angles θ and Φ.

Initial component	Final components		
I_x	$I_x(1-\cos\alpha)\times$ $\times\sin^2\theta\cos^2\Phi+$ $+I_x\cos\alpha$	$I_y(1-\cos\alpha)\times$ $\times\sin^2\theta\cos\Phi\sin\Phi+$ $+I_y\sin\alpha\cos\theta$	$I_z(1-\cos\alpha)\times$ $\times\sin\theta\cos\theta\cos\Phi-$ $-I_z\sin\alpha\sin\theta\sin\Phi$
I_y	$I_x(1-\cos\alpha)\times$ $\times\sin^2\theta\cos\Phi\sin\Phi-$ $-I_x\sin\alpha\cos\theta$	$I_y(1-\cos\alpha)\times$ $\times\sin^2\theta\sin^2\Phi+$ $+I_y\cos\alpha$	$I_z(1-\cos\alpha)\times$ $\times\sin\theta\cos\theta\sin\Phi+$ $+I_z\sin\alpha\sin\theta\cos\Phi$
I_z	$I_x(1-\cos\alpha)\times$ $\times\sin\theta\cos\theta\cos\Phi+$ $+I_x\sin\alpha\sin\theta\sin\Phi$	$I_y(1-\cos\alpha)\times$ $\times\sin\theta\cos\theta\sin\Phi-$ $-I_y\sin\alpha\sin\theta\cos\Phi$	$I_z(1-\cos\alpha)\times$ $\times\cos^2\theta+$ $+I_z\cos\alpha$

In Table M.2 we consider the particular case of an on-resonance pulse ($\theta = 90^\circ$).

Table M.2. Final Angular Momentum Components after Rotation by an Angle α about an Axis in the xy Plane (θ=90°) Defined by Angle Φ.

Initial component	Final components		
I_x	$I_x \cos^2\Phi +$ $+I_x \cos\alpha \sin^2\Phi$	$I_y (1-\cos\alpha)\times$ $\times\cos\Phi\sin\Phi$	$-I_z \sin\alpha\sin\Phi$
I_y	$I_x (1-\cos\alpha)\times$ $\times\cos\Phi\sin\Phi$	$I_y \sin^2\Phi +$ $+I_y \cos\alpha\cos^2\Phi$	$I_z \sin\alpha\cos\Phi$
I_z	$I_x \sin\alpha\sin\Phi$	$-I_y \sin\alpha\cos\Phi$	$I_z \cos\alpha$

Very often the phase angle Φ can only take one of the "cardinal" values 0°, 90°, 180°, 270°. In this case, the product $\sin\Phi\cos\Phi$ vanishes and the expressions above become even simpler, as shown in Table M.3.

Table M.3. Final Angular Momentum Components after Rotation by an Angle α about an Axis in the xy Plane (θ=90°) with the Angle Φ a Multiple of 90°.

Initial component	Final components		
I_x	$I_x\cos^2\Phi +$ $+I_x\cos\alpha\sin^2\Phi$	0	$-I_z\sin\alpha\sin\Phi$
I_y	0	$I_y\sin^2\Phi +$ $+I_y\cos\alpha\cos^2\Phi$	$I_z\sin\alpha\cos\Phi$
I_z	$I_x\sin\alpha\sin\Phi$	$-I_y\sin\alpha\cos\Phi$	$I_z\cos\alpha$

Making Φ equal to 0°, 90°, 180°, 270°, in Table M.3, yields the rotation rules about *Ox*, *Oy*, – *Ox*, – *Oy* (phase cycling). It is instructive to compare these results with those obtained by means of the vector representation.

Suggested Readings

R.R. Ernst, G.Bodenhausen and A. Wokaun, *Principles of Nuclear Magnetic Resonance in One and Two Dimensions*, Clarendon Press, Oxford, 1987.

Ray Freeman, *A Handbook of Nuclear Magnetic Resonance,* Longman Scientific and Technical Publishing Co., Essex, 1988.

O.W. Sörensen, G. Eich, M. Levitt, G. Bodenhausen and R.R. Ernst, *Progress in NMR Spectroscopy*, **16**, 1983, 163-192.

Ad Bax, *Two-Dimensional Nuclear Mgnetic Resonance in Liquids*, Reidel Publishing Co., Dordrecht/Boston/Lancaster, 1982.

T.C. Farrar and J.E. Harriman, *Density Matrix Theory and Its Applications in Spectroscopy*, Farragut Press, Madison, WI, 1991.

M. Goldman, *Quantum Description of High Resolution NMR in Liquids,* Clarendon Press, Oxford, 1988.

C. P. Slichter, *Principles of Magnetic Resonance*, Springer-Verlag, Berlin/Heidelberg/New York, 1980.

G.E. Martin and A.S. Zektzer, *Two-Dimensional NMR Methods for Establishing Molecular Connectivity: A Chemist's Guide to Experiment Selection, Performance and Interpretation*, VCH Publishers, Weinheim/New York, 1988.

G. D. Mateescu and A. Valeriu, "Teaching the New NMR: A Computer-Aided Introduction to the Density Matrix Formalism of Multipulse Sequences," in *Spectroscopy of Biological Molecules,* Camille Sandorfy and Theophile Theophanides (Eds.), Reidel Publishing Co., Dordrecht/Boston/Lancaster, 1984, 213-256.

G. D. Mateescu and A. Valeriu, "Teaching the New NMR: A Computer-Aided Introduction to the Density Matrix Treatment of Double-Quantum Spectrometry," in *Magnetic Resonance,* Leonidas Petrakis and Jacques Fraissard (Eds.), Reidel Publishing Co., Dordrecht/Boston/Lancaster, 1984, 501-524.

INDEX

P

Q

R

S

T

V

W